ary STORY

A Dairy Story

By David and Wilma Finlay

First published in 2022 by
Finlay's Farm Limited
Gatehouse of Fleet, Castle Douglas, Scotland, DG7 2DR
www.TheEthicalDairy.co.uk

ISBN: 978-1-7391039-2-7

Copyright © David Finlay and Wilma Finlay 2022
The authors have asserted their right under the Copyright,
Designs and Patents Act, 1988, to be identified as the Authors of this Work.

All rights reserved. No part of this publication may be reproduced,
stored or transmitted in any form, or by any means (electronic, mechanical,
photocopying, recording or otherwise) without the prior written permission
of the publisher or authors.

A catalogue record for this book is available on request
from the British Library.

Illustrations copyright © Ian Findlay
Typesetting by Findlay Design

Distributed under license by
5m Books Ltd
Lings, Great Easton
Essex CM6 2HH, UK
Tel: +44 (0)330 1333 580
www.5mbooks.com
Printed and bound by CPI Group (UK) Ltd, Croydon, CR0 4YY

To everyone who visited the farm over
the last 25 years and asked,

'Why don't you keep the calves with their mothers?'

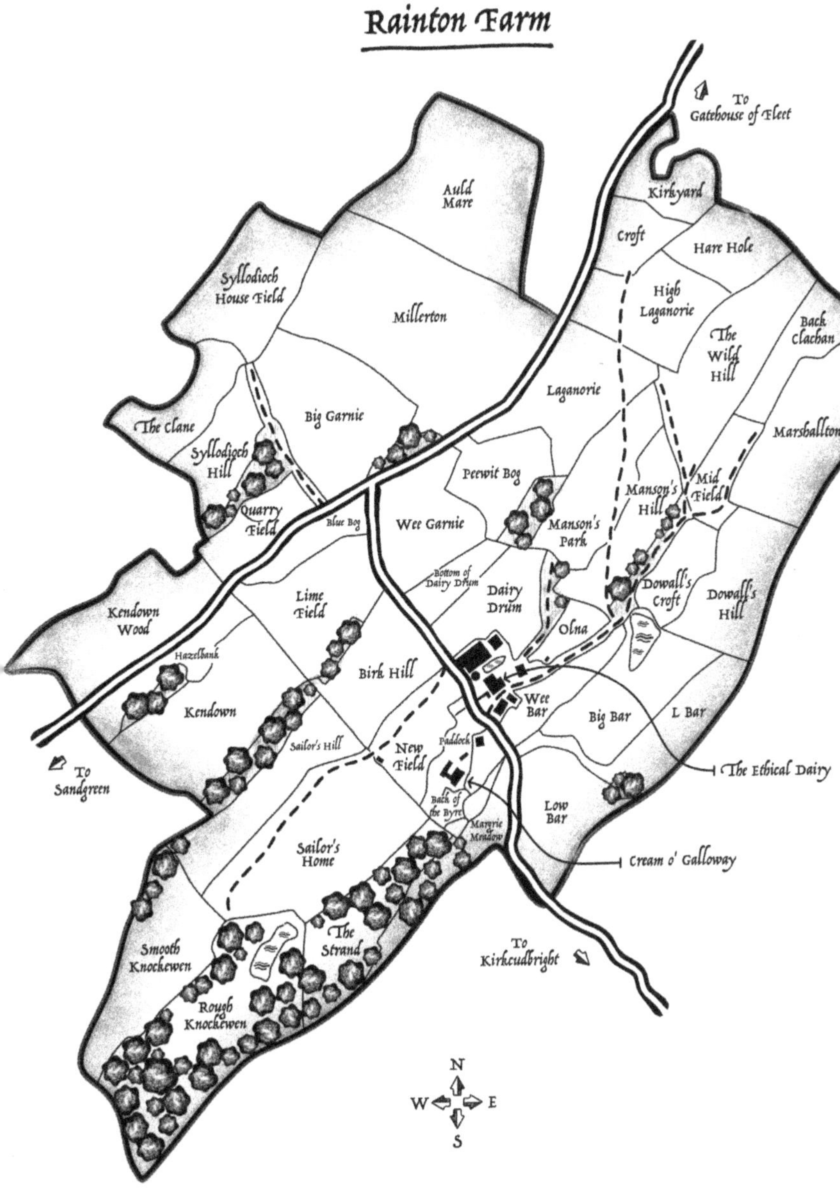
Rainton Farm
To Gatehouse of Fleet
Kirkyard
Auld Mare
Croft
Hare Hole
Syllodioch House Field
Millerton
High Laganorie
Back Clachan
The Wild Hill
Laganorie
Big Garnie
The Clane
Marshallton
Syllodioch Hill
Peewit Bog
Quarry Field
Blue Bog
Wee Garnie
Manson's Park
Manson's Hill
Mid Field
Bottom of Dairy Drum
Dairy Drum
Dowall's Croft
Dowall's Hill
Kendown Wood
Lime Field
Olna
Hazelbank
Birk Hill
Kendown
Wee Bar
Big Bar
L Bar
Sailor's Hill
Paddock
New Field
The Ethical Dairy
To Sandgreen
Back of the Byre
Low Bar
Margrie Meadow
Sailor's Home
Cream o' Galloway
Smooth Knockewen
The Strand
To Kirkcudbright
Rough Knockewen
N
W
E
S

Introduction

By Wilma Finlay

David woke me up at 3am and said 'That's it, we're giving up'. I didn't need to ask what he meant because it was all we had been thinking about for weeks. We couldn't actually talk about it without arguing. Our attempt at keeping the dairy calves with their mothers had failed.

The stillness of our bedroom on that crisp March night in 2013 was in stark contrast to the crushing chaos of the dairy shed, just a few minutes' walk from the farmhouse. With the inevitable decision now made, a weight lifted off my shoulders as my heart broke.

For the previous five years we had been working towards a complete transformation of our dairy farming system. Our goal was to put an end to the standard industry practice of removing dairy calves from their mothers soon after birth.

We wanted to prove to a cynical industry that it was possible to do dairy differently; to design a kinder method of dairy farming – a system where the distressing, heart wrenching bawling of mothers forcibly separated from their newly born calves need no longer be heard.

Dairy cows produce a lot more milk than a calf needs, so we designed a farming system where we would share the milk with the calf, moving from twice-a-day milking to once-a-day, and letting the calf stay with its mother. We had sunk everything we had and more into this grand experiment – emotionally, financially and, worse than that, we had done all this very publicly. But our new system didn't work.

The calves were drinking almost all of the milk their mothers were producing, which meant we had virtually no milk to sell. Our original plan had been to keep the dairy cows and their calves together until the cow was close to calving again. At that point, we had figured, the cow would be losing interest in her calf and getting ready for the birth of her next one. Meanwhile the calf, at almost a year old, would be spending most of its time playing with its mates rather than needing attention, or milk, from its mum. But even at this stage, with the calves just four months old, this system was bankrupting us. You only had to lift the lid of the milk tank to see there was hardly enough milk to make ice-cream, far less make cheese. The monthly milk cheque that the farm relied on as its main source of income was laughably small.

David went through all sorts of scenarios as to how he could keep it going. The most delusional was that he'd milk the cows 365 days of the year to save on staff costs. Considering he was already working 70 hour weeks, this was not a realistic option. We started to talk about how we would begin the separation of the cows and calves. We couldn't just stop immediately and make the calves go 'cold turkey'. David came up with a plan to separate them overnight and then, after milking the cows first thing in the morning, we'd re-unite them with their calves for the rest of the day. 'You never know, that might actually work,' he said more in hope than expectation.

...

With the benefit of hindsight, we now understand the importance of this painful experience in helping us design a cow-with-calf dairy farming system that could work, but there was nothing inevitable about it at the time. Cow-with-calf dairy farming was virtually unheard of back then, and it had certainly never been done at scale in a manner that allowed both the mother and her calf to thrive. The early pioneers of cow-with-calf dairy were farming at a micro scale, a handful of cows and a small volume of milk that was easy to sell locally. We were trying to do something very different, with more than 100 cows and half a million litres of milk.

Our farm, Rainton, is a family-farm sized enterprise in south west Scotland, in the very heart of Scotland's industrial dairy country. It's the type of dairy farm that's being squeezed out of existence as intensive 'big dairy' expands ever onwards, removing dairy cows from green pastures and enclosing them in indoor mega-sheds 365 days of the year.

David and I firmly believed that intensification shouldn't be the inevitable future of a farming sector his family had been part of for generations. We wanted to find another way. Now, as we look out at the cows and calves contentedly grazing the grasses and herbs in our fields, we know without a doubt that there is, but finding that alternative path wasn't easy.

A comment I've noticed cropping up on our social media posts recently is 'why can't everyone farm this way?' This book is an answer, of sorts, to that question, but we hope it is more than that. Our planet stands at a tipping point of ecological and climate catastrophe. How we farm at Rainton today isn't just about keeping dairy cows and their calves together, it's about acknowledging that nature knows what it's doing a damn sight better than humans do.

The journey that David and I have been on together spans thirty years, and it's taken fifteen of those years to get to the point where we can say, with confidence, that this system works. It works for the cows and calves, it works for us and our team, it works for our customers, it works for wildlife and it works for the planet. We think it can work for other family-sized farms too.

David and I were born within a few months of each other. We are now both in our late 60s. While David's family is long-lived, cancer has been in the background for much of my adult life, coming to the fore every now and then. It's here once again, and this time it's not going away. We understand what this means for our future. For that reason, and many more, now feels like a good time to write about our journey and what it has taught us.

Farming needs to change, and it needs to change quickly. The most important thing we have learned is that change is possible, and that in itself gives us hope. Our experience has proven to us that the transformation of complex eco-systems for the benefit of planet, animals and people is not only possible, it's beautiful and it's immensely rewarding.

But we also know that change isn't easy, and that transformation takes time. The need for transformation across so many systems is frighteningly urgent. We hope that sharing our experience might help to show how system change can happen, what it takes as well as what can help to make it all work.

So, this is our story.

Chapter One

David

As a youngster dragged out endlessly in cold, sleety weather to help with the lambing, calving, stock-work and so much more, I swore I'd never be a farmer. My resolve held until my early thirties. With my father retiring I was forced to decide whether to continue agricultural consultancy with good pay and conditions, or return to practical farming with long hours and barely enough income for me and my family to live on. I chose the latter. So began an unexpected farming journey, stepping off the well-trodden path towards intensification, and embarking on an expedition into the unknown.

Few of us give more than a passing thought to the stuff we put in our mouths, yet it touches every aspect of our lives – from our health and wellbeing, to our very survival as a species. For much of my life I was no different. It was a gradual, growing disillusionment with the side-effects of intensifying our farming practices, exposure to public feedback, and concern about the climate and environment, that conspired to set us on a course of questioning the standard food-industry mantra. But asking questions that challenge the fundamental principles of such a traditional industry is neither easy nor welcome.

When you're a sixth generation dairy farmer, and distantly related to probably half the local farming community, there are acceptable variations of farming practice and there are unacceptable ones. Acceptability is narrowly defined to be what everyone else is doing – driven by history, culture, peer pressure, education and industry agendas. As farmers we grow up together, learn together, work together, socialise together and we tend to live in our own little 'country-ways' bubble.

Innovation in farming tends to be defined as applying the latest technology that promises to extract an extra litre of milk, kilo of beef or corn, carry an extra cow or two, cut out a worker or reduce some other cost – basically, doing more of the same but cheaper, faster and in more volume. In other words, intensification. To question this convention is seen as a criticism and can be considered offensive. To do so as a 'family' member is borderline traitorous.

...

The Finlays moved down to Galloway from Ayrshire in the late 1700s. Farming folk who came out of Ayrshire understood poverty, and so they knew the value of money. The saying was they'd go from rags to riches and back to rags again in just three generations.

My ancestors took on a smallholding near to where we are now and they struggled. The land is stony and rocky here, next to the rugged edge of south west Scotland, and it would have been a hard toil back then. In many ways, it still is. An opportunity arose to manage the local distillery, called The Stell, in nearby Kirkcudbright, and my great, great grandfather John jumped at the chance. Apparently, and unsurprisingly, his new position garnered a great many friends, but still no money.

In the mid-1800s the railways rolled out across the countryside, ribbons of connections linking small communities like ours with the rest of the world for the very first time. Before this the roads had been poor – treacherous even – and much of the rural trade was done by sea rather than by land. Small ships would lay-off the various beaches and coves as traders rowed out to secure supplies of coal, sugar and more. Stories of smugglers evading excise men along this craggy coastline are legendary. Poet Robert Burns was one of those in hot pursuit of contraband, another dairy farmer from Ayrshire whose farming ambitions were frustrated by unproductive soil and soul destroying hard work.

The arrival of trains offered estate owners the opportunity to add value to the milk produced here, by building dairies with cheesemaking facilities attached. Cheese would be made over the summer months and sent by horse and cart to the local train station, then onwards to the cities. Fresh milk wouldn't travel well back then, but cheese did.

It seems hard to believe but even in the late 1800s the fresh milk consumed in urban areas was produced in intensive dairy farms located right in the heart of cities. Up to a thousand cows would be housed in enormous byres, with all the feed carted in and the muck carted out. All that remains of these city dairies are the names left behind, think Byres Road and Cowcaddens in Glasgow, for example. I find the stories of these dairies fascinating, and the parallels between the city dairies of the industrial revolution and the intensive dairies of today are not lost on me. What was different back then was how labour intensive milking was. It took 100 women to keep a thousand cows milked twice a day. Nowadays big dairies often do all their milking using robots.

In the 1860s my great, great grandfather John got the chance of leaving the distillery to become a tenant on a 60-cow dairy and cheese farm near the village of Borgue. Just a few years later, in the 1880s, the modern milking machine was invented saving dairymen from the arduous chore of hand milking. That milking machine, created by Stewart Nicolson of Bombie Farm, Kirkcudbright, just a few miles from here, transformed dairy farming around the world.

By the 1930s times were particularly hard. The Great Depression saw my grandfather take on a second tenancy, this time at Rainton Farm – the farm where we now live. This wasn't unusual for the time. He had two sons and it was considered correct for a farmer to provide the opportunity for his sons to have an equal chance of farming. My father took on the running of this farm after returning from active service in the Second World War. He would travel by motorbike from his home seven miles away until he married my mother in 1952, when he moved into the semi-detached farmhouse here at Rainton.

The farm was a rugged 650 acres with rocky, scrub-covered outcrops in most fields, often with boggy hollows. The soils are of a kind, loamy brown earth, but very stony. It's about a mile from the Solway coast, generally south-west facing and it rises from 30 to 300 feet above sea level. Over the decades my father and the farm team laboured to – as he said – 'turn it from a third-rate farm into a second-rate one.' For me, it was home.

Childhood on the farm was amazing. There were enough kids to form two football teams and half-fill the school bus. I was the second of six children in my family, but there were also the four kids of the McCarlies, three of the Whites, six of the Hunter family, and five McKnight children. We pretty much ran wild. There were forts built from hay bales, and we burrowed tunnels, just wide enough to squeeze through, into the loose straw in the old barn. We made rickety, home-made go-karts and careered down the farm road at speed. As the nights darkened there was kick-the-can, hide-and-seek and sardines to keep us entertained. The snowy winters of the 1960s brought snowball fights, group sledging on sheets of corrugated tin, igloo-building and snow tunnels. I'm still amazed none of us were killed. Plenty of bumps and bruises fairly toughened us up.

By the time I was ten, I was old enough to help with lambing. This wasn't the sanitised version of lambing that's shown on TV, it was outdoors in the cold, wet weather of March in Scotland. Slipping and slithering through muddy puddles, dangling a lamb in each hand as the sleety showers froze fingers stiff while the infuriated mother sheep followed, either head-butting you or running off into the distance.

There were turnips to hand-load into trailers and take out to the fields for the newly lambed ewes and lambs, and we hand-forked the turnips back off the trailer. If you were lucky, someone drove the tractor while you forked, but often there was no-one available. I came up with my own way of tackling this particular job. I'd put the tractor into low gear, jump off and run round the back, climb onto the trailer, fork like mad then run back to the tractor. The trick was to avoid being crushed by the rear wheel, changing direction before tractor, trailer and I careered over one of the many steep-sided, gorse-covered banks.

Then there was stone-picking. We reckoned a day's ploughing created a week's worth of stone-picking. There was the 'first-pick' which were the biggest stones that might damage the harrows. Then, before sowing oats or turnips, we did the 'second-pick' – removing the numerous smaller, fist-sized stones. Somehow, no matter how many stones we picked, the next time the field was ploughed their stone friends had turned up to taunt us all over again.

Early spring drifted into late spring. This was fledging time for the cacophony of rooks and jackdaws that nested in the nearby woods. As children we knew that jackdaws were the enemy. They dug out turnip seedlings searching for leatherjackets – the juicy grub of the 'daddy-longlegs' cranefly. Then they would arrive in flocks in the autumn and knock down the oats to eat the grain.

They were bad! Our job, from around twelve years of age, was to go into the woods with our 410 shotguns and blast the fledgling crows as they balanced on the edge of their nests preparing to fly. Once they were flying it was too late, they were far too smart to let you get near them once they took to the air.

Another job given to us children was to take several trays of hens' eggs round the farm. In those days the farm game-shoots were a big part of the social calendar. Pheasants and partridges bred naturally on the farms and anything that might take a ground-nesting bird's eggs or chicks was bad. This meant foxes, crows, and all birds-of-prey were also the enemy. Badgers, buzzards and red kites didn't exist in this area in those days, they had been virtually exterminated. I didn't even know, until many years later, that they were indigenous. All these animals and birds were shot, trapped and poisoned. That was the purpose of the hens' eggs. There was a little hole in the top of the egg and sometimes a spot of green deposit. Our job as children was to make little nests with a couple of poisoned eggs all around the farm.

After the turnips, came hoeing. With luck there might be a team of hoers. This was quite enjoyable, if a bit wearying by the end of a hot day in May. It was an opportunity to hear all the local gossip and stories, most of which were invariably greatly enhanced.

Spring rolled into summer and next in the series of farm kid entertainments was dagging. The fresh grass growth would make the poo of the ewes and lambs a bit runny, which would stick to the wool around their tails. This could be very dangerous because it attracts blowfly. The blowfly lays thousands of eggs in dirty wool which then hatch into tiny maggots. At first, they feed on the poo and grow rapidly. Then they turn their attention to the lamb. Dagging was a case of holding the lamb firmly between your knees and, doubled over, snipping off the dirty wool. It was back-breaking work but lifesaving for the lambs.

Next in our endless list of seasonal chores was haymaking. How we ever made decent hay in the west of Scotland defies me. Over the course of the summer we made many hundreds of small round bales. A few days after baling, our job was to kick the bales over 180 degrees to stop them drawing damp, to stop the fresh grass from growing into them, and to tie-off the loosely wound string that held them together.

In those days we would have a fortnight's summer holiday with our cousins at our local, beach-side caravan site a mile away. That was an adventure and it was really the only time we spent off the farm, but even then, farm duties often called, heralded by my father's appearance at 7am some mornings drumming us up to, 'come roll bales'.

Unquestionably Dad was a big factor in my early development. He'd joined the army in 1944 after leaving college, just in time to take part in the murderous crossing of the Rhine in the final weeks of the Second World War. He didn't need to go. The eldest sons of farmers could opt to stay home and produce vital food to feed a hungry nation, but if he hadn't gone it would have meant his younger brother would have been drafted, so he went.

Like most of the survivors he never talked about it. It was only when war history researchers, both British and German, contacted him on the 50th anniversary of the end of the war that we began to get a glimpse of what he had gone through. What all those young men had gone through.

Dad's limp was a visible sign of his war experience, after being wounded in the leg with a smashed ankle. In some ways, I guess, he was lucky. The injury meant his war was over. It also taught him that you had to be tough to survive, and that was the approach he applied to raising his kids. Tough love. With his damaged ankle he found walking any distance a problem, so he cycled everywhere. I remember as an infant being perched side-saddle on his bicycle bar, which would have an old jute sack wound round it. I recall clinging on to the inner part of the handlebars for dear life as he cycled furiously over rough ground from field to field to tend stock.

In contrast, there was Mum, and her influence was one of compassion and fun. She laboured day and night scrubbing, cooking and tending the garden for a husband and six kids. We weren't poor by the standards of those days, but every penny counted. Clothes were handed down, casualty livestock were shot and butchered on the farm. My abiding dread was mutton hot-pot, with strong sheep-smelling liquid fat floating on top. No food was wasted. My parents had come through a war where people knew real hunger, so we sat at the table until our plates were clean.

My mother's fun side shone through on car journeys where we sang popular songs. On hot summer afternoons she'd arrive at the school gates in our old Humber car, pick us all up and take us to the beach for a swim. One of her comments to me that has stuck over the years is, 'David, you take life far too seriously!' I probably do. Her compassion for all forms of life, from butterflies and bees to abandoned baby hedgehogs or hatchlings, was tested each year at lambing time when new-born lambs were trailed into the farmhouse near the point of death. It would take a massive effort to nurse them back to health, but she did so with gentleness and care. Mum never struck us, that was Dad's job, with one memorable exception.

I might have been five or six. I walked in and Mum and some of the younger kids had built an impressive tower with wooden play-bricks. Don't ask me why, but I kicked over the tower which collapsed. She smacked me – hard – and ordered me to bed. As I lay there, tearful and resentful, Mum came to the room door and softly begged me to forgive her. To my eternal shame I resisted the powerful temptation to turn and give her a hug.

Nevertheless, at an early age I was beginning to sense that I wasn't exhibiting the behaviour that was normally expected of a young man. When the 'big boys' at school were playing 'knifey' with a poor, unfortunate toad they'd found, I walked away feeling nauseated.

When I was about 11, Dad was showing me how to use the .22 rifle. He picked a target on a nearby tree which I was horrified to see was a song-thrush. It was far enough away that he'd probably miss, I thought. He didn't. He handed me the gun and I left quickly before he noticed the tears. I really didn't want to shoot anything, except, maybe, crows because they were bad. Oh, and rabbits. There were thousands of them, and they were bad too.

I took part in the farm game-shoots, of course. My poor success rate was put down to being left-handed but with a dominant right eye. I did enjoy the shoots because of the craic and getting to see other farms, but my lack of killer instinct was eventually 'outed' when, in full view of the shooting party, I stood and watched as a roe-deer doe ran past me within just a few yards and I didn't even raise the loaded shotgun. The gamekeeper was apoplectic. That was my last farm shoot.

At the age of twelve my father used some inheritance money to bundle me off to boarding school in Edinburgh. I was naively relishing the prospect of the new experience, getting away from the farm and, more importantly, farm work. Most of the boys seemed to enjoy boarding school, but I hated it from the moment I arrived until the moment I left. Occasionally a boy would run away and while I never did, I was sorely tempted. Maybe it was the realisation that my parents were sacrificing a lot to send me there. Even years after leaving school, when approaching Edinburgh by car I'd often find a knot of tension building in the pit of my stomach. Undoubtedly boarding school achieved what my father intended. Links to an elite network – which never interested me – a break with the local 'group think' and a sort of weaning from the comfort of family. It certainly toughened me up.

Time at home was never easy. My father and I rarely got on. There was an occasion when, after a row, I packed a rucksack and walked out to the A75 main road. I hitched a lift and was on my way. Problem was, to where? I was 16 and really didn't know anyone outside the local area.

I remembered a couple of music teachers with two young children who had come to the farm a few years earlier and stayed in a semi-derelict cottage for holidays. They had said, 'Come down to visit, any time.' I knew the town where they lived in Somerset, but it was 380 miles away. They got a bit of a shock to find me sitting on the grass verge outside their house at 8 o'clock the following morning. They were brilliant about it and I stayed a couple of weeks. I guess they phoned my folks. I didn't ask.

Next step was university. I'd briefly dabbled with the idea that I might not go down the farming route, and even took a train trip down to Aberystwyth to check out a course in marine biology, but inertia prevailed, and I predictably got a place at Aberdeen studying agriculture. In my first year, as the agri-course wasn't too challenging, I also signed on for maths. I enjoyed maths but had done badly at school, perhaps because our maths teacher was an alcoholic.

Studying agriculture bored me and towards the end of the first year I applied to Glasgow Vet School on the off-chance they might have an opening, and I got an interview date for mid-summer. Early that summer I went hitch-hiking round Europe. It's hard to believe how popular hitch-hiking was in those days. There could be half-a-dozen, or more, young folk strung out along a motorway slip-road 'thumbing it'. I travelled across France, down into Spain, back north to Germany and then through the Netherlands to France and back to the UK. Two months of exploring, lost two stone, spent thirty quid, met a fascinating variety of colourful, friendly and, sometimes, scary people. I even worked as the roadie and lights engineer for a touring French rock group. It was tempting to stay with them, but I had an interview to get back for. I travelled up to Glasgow in late August and found my way to the interview room in good time. That must be worth a few points, I thought. It didn't cut much ice as the panel had misread my school maths grade as a 'C', when in fact it was an 'E'. It was a short and testy interview.

It was well into my second year at uni that a vivacious young woman in our year turned her attentions to me. I'd been watching her from a distance from day one but realised Jane was on a different planet as far as social relationships went. I was a quiet introvert, she wasn't. They say opposites attract and I don't doubt that. My straw-poll suggests, however, that they don't last. To say our relationship was turbulent would be an understatement, and how we managed to raise three brilliant kids defies me. Undoubtedly, they were the cement that held it all together for as long as it did and they gave me the motivation to make something more of my life than just drifting back home.

A bare minimum honours degree secured me a job with the Agricultural College as a farm consultant in Banffshire, in north east Scotland. We were married with an eighteen month old son, Mark, and a second child on the way when we moved near Turriff, living in a small estate cottage. I was 22 and life was good.

Banffshire is a delightful county of mixed arable and livestock farming and the folks are warm and welcoming. Our second and third children, Margaret and Christine, were born there and I'd have stayed longer but for a change of office boss that brought unbearable friction, and an opportunity to run the farm consulting office in Shetland. Jane was up for it and the guys in the office joked about how little there would be to do in Shetland. They couldn't have been more wrong.

We moved into our council house in the village of Voe in October, just as the long, wild Shetland winter also moved in. The culture shock was palpable. This was 1981, I was 26 and the Sullom Voe oil terminal had just been completed. Shetland was awash with oil, oil-money and politics.

My wake-up call came early that first winter. It was an evening meeting held by the Shetland Islands Council where the council's head of finance was explaining their new agricultural strategy that would use some of the oil revenues to provide cheap development loans to crofters and farmers. There was discontent among the large crofter audience because they'd have preferred grants, but the council guy assured them that I, as the agricultural expert, had been consulted and supported the plans. I was at the back of the hall sitting next to a crofter who I subsequently became great friends with. He looked at me and raised his eyebrows. I made it clear that I knew nothing. He was the farmers' union secretary and stood up to point out to the whole room that I had something to say.

I explained to the audience that I knew nothing of the council's plans, and that this was the first I'd seen of them. The meeting was drawn to a close and, as the last of us were leaving, the council's Chief Executive strode over and asked to have a 'quiet word'. We went into a side room where he grabbed me by my jacket lapels, shoved me back against the wall and hissed, 'If you ever contradict one of my officers in public again, I'll personally make certain you never work again!' I was utterly flabbergasted. Nothing in my training had prepared me for this.

There were three of us in the office, Graham a steady young graduate and a farmer's son, Stella our office manager and me. Stella was in her late fifties. She'd been the daughter of the school-teacher on one of the most remote islands in the UK – Foula – before moving to a croft on another island – Bressay – near the island capital of Lerwick where she raised her own family. Stella was intelligent and wrote poetry, and she was my window into the real Shetland. She knew everyone, and knew who they were related to going back several generations. She also told me what I should be careful of.

Stella had been raised during the time before electricity reached the remoter islands and she recounted life in those days. She explained how in her youth the common folk in Shetland had to doff their caps to the power-players – the landowners and clergy – but now, with floods of oil money pouring into the council coffers, that elite role was filled by councillors and their senior officials. That explained a lot.

Shetland folk are very dear to my heart and I have many fond memories of my time there. Our work involved putting together development plans for the crofters and farmers, funded by council loans. It also involved working with BP, who managed the oil terminal, on their reinstatement programmes for the construction camps and pipeline tracks. Then there was the Council and their various infrastructure projects. There was the Department of Agriculture, the Crofters' Commission, Highlands and Islands Development Board, Scottish Natural Heritage, the RSPB and others drawn to Shetland by the oil. Committee work was endless and, in the end, largely fruitless.

There was one committee that brought me into direct contact with 'the enemy' – the Farm and Wildlife Advisory Group. Half the members were farmers or crofters, the other half were environmentalists. At first it was daggers drawn, but as the months and years rolled by, common ground emerged and a much greater understanding and mutual respect developed. Up to this point I'd seen environmental stuff as an obstacle on the road to agricultural development, and I was clear that organic was a joke, a fraud. This group was the start of my journey towards a different mindset.

Chapter Two

Wilma

I grew up with a chip on my shoulder – it's possibly still there. My father worked long hours as a painter and decorator, but like most families in the 1960s in the north east of Scotland, we had little money. My sister and I did well at school, but many of our classmates, friends and relatives were from professional backgrounds. While most of them went on school cruises and outward bound trips, we knew better than to even ask our parents if we could go too.

I'm definitely competitive, and the only way I could compete as a child was academically. I was at school in the days of streaming, and we had classes A to F. I was in the A stream and most of my friends were in the A and B classes. Looking back now, none of my classmates were farmers' sons. I was friendly with farmers' daughters but farmers' sons didn't exist in my life, so I wrote them off as being uninteresting and not that bright.

Both my father and mother's parents had small farms. By the time I came along my mother's parents had retired to a council house in a nearby village and had hens in their garden, while my father's parents had retired to a smallholding with a milking cow and hens. To me they were the hub of the neighbourhood. The postie nearly always had a cuppa with them, and the neighbouring farmer helped them make hay and do repair jobs.

We lived five miles away and during the school summer holidays I loved cycling to visit my grandparents to help out; milking the cow by hand and feeding the hens. It was my granny who taught me to cook and bake. All her recipes were 'a handful of this and a wee picky that', which no doubt explains my inconsistent baking to this day. There was always something to do and sitting down for a rest usually coincided with neighbours popping in to share recent gossip. My aunt often compared me to my grandmother – always had to be busy, always trying something new.

We lived in a rural part of Morayshire, between Inverness and Aberdeen. The nearest village was a mile away, and the only other children in the immediate neighbourhood were the four Rose boys who lived next door. They certainly weren't going to play with our dolls, so my sister, Heather, and I played football with them. We were real tomboys.

When she was fifteen, Heather joined the local girls' football team. I was so jealous. I was only twelve, but I blagged my way into the training sessions and got to play in the team. We were called the 'Forres Flamingos' and we went for a long run of games without losing, including a game against the Aberdeen Prima Donnas at Pittodrie. My abiding memory is playing in a cup final and missing a sitter of an open goal. We only drew that night and I was dropped for the replay. The disappointment was devastating, but the real heartache came when the team was disbanded a couple of years later, when some of the best players all left to go to college at the same time. I sobbed. I've still got my Flamingos strip.

It's probably only when you're older that you really appreciate your parents. While I'm now proud to say we had a normal, stable family life, as a bolshie teenager I thought my parents were boring. Dad had fought and been injured in the war, but he never talked about it. We knew he'd spent some time in hospital recuperating because my sister and I each had a stuffed toy kangaroo that he'd made during his rehabilitation. I guess we all have something we're ashamed of about things we said to our parents when we were young. I never asked, 'What did you do in the war, Dad?' But during an argument I did say, 'What do you know about the world anyway?' I still remember his derisory laugh. At that time I didn't know about his training in Canada, or that he'd been a prisoner of war in Germany, or that he'd been shot in the mouth in Italy. The bullet had gone in one cheek and out the other. The plastic surgeons had done a remarkable job of patching him up.

Mum put everyone before herself. She was a very meek woman but extremely bright, having been the Dux of her school. Despite her clear ability her parents' expectations of her were modest; that she'd get married, have children and look after the family. That did happen – eventually – but due to the Second World War she was 33 before she married. In the meantime, she attended secretarial college and worked in a large estate office. Like most woman in the 1950s, she gave up her job when she married, and never worked outside the home again. I knew that I didn't want the same role my mother had. I was absolutely determined that I would always be financially independent, no matter what my personal circumstances were.

In a small town in rural Scotland back then everybody knew everybody else. And people didn't just know your parents and siblings, they knew both sets of grandparents as well, and you were judged by the family package. As a typical awkward adolescent this was suffocating to me, and I was more than ready to escape.

I studied Mathematics at Edinburgh University, and the choices of subject and university were easy decisions. It had nothing to do with future career choices or anything like that. It was simply that maths was by far my best subject at school, and Edinburgh was far enough away from home that I would be able to be anonymous. The first three years of my degree were everything I could have hoped for. Great flatmates and parties and, of course, some studying to make sure I got through to the next year. The final year was hard work – really hard work. It was the first time I started to worry about exam results, realising that whatever degree I achieved could have a major impact on my future.

Apart from the super-brainy, we were all in the same boat, so it was a solid nine months of studying. I remember the shock when, just a month before finals, one classmate decided to quit the course entirely. 'It isn't worth the stress,' he said, yet he didn't appear to be someone overly bothered by stress. I couldn't believe he was going to throw away the degree he had worked towards for years, for the sake of one month of study. But then again, he was a farmer's son and he knew his career didn't require an honours maths degree.

I get quite cynical about university courses now. University is a wonderful stepping stone into adulthood, which I don't regret for one second, but who really needs to understand sixth degree polynomials or the wonderfully named Hairy Ball Theorem. In all my working life I've only used addition, subtraction, multiplication and division and, of course, logic. Just think how much stress we'd remove from teenagers if we just gave them enough maths knowledge to calculate percentages and understand basic statistics.

I achieved a respectable degree and it was time to find a job. This was an era when a degree would open doors, and I swithered between getting a job and doing a PhD. I opted for an Industrial PhD at Aston University in Birmingham, where I worked for a company in Kidderminster and had a weekly class with my tutor at university. Boy, was work a shock! I was in an office where no one enjoyed their job. The week revolved around football and football-related betting. Monday – discuss the football results; Tuesday – do Spot the Ball; Wednesday – discuss Spot the Ball results; Thursday – debate the pools; Friday do the pools. It was a different world, and it was the first and only time that I ever skived.

I should have been there for three years, but I only lasted 18 months. I didn't finish my PhD, but it wasn't without value, because it introduced me to Comshare, a computer company that had offices all over the UK, and they were looking for staff in Glasgow.

There couldn't have been a greater contrast. The Comshare team in Glasgow was full of young people, enthusiastic about their jobs and working hard. This was the early 1980s and computers promised to revolutionise the world. My company was in expansion mode and every three months there was a four week long induction course for new starts. It was an adrenalin rush from beginning to end, with friends made for life. The hours were long and the food and drink flowed. It was a very steep learning curve, but I thrived on it.

After three years, I was asked to relocate to the Aberdeen office. The oil industry was booming and the computer systems within the oil companies were not coping with the speed of change – they needed urgent external support. After Glasgow, the social life in Aberdeen was a bit of a disappointment. It was either the 'oilies' or people who had settled down. So I did what I knew best – worked – although I did have one release – football. I was in Aberdeen during the Alex Ferguson era and Pittodrie was my home every second Saturday afternoon. I will never forget watching Aberdeen beating Bayern Munich on their way to becoming European Cup Winner's Cup Champions. One of the reasons I will always remember that game is because the following morning I flew to Munich for a job interview.

The lack of social life in Aberdeen had been really getting to me, so when a former Comshare colleague phoned me with the opening words of 'Are you ready for a proper job yet?' I found that I was. He'd left to join Motorola in East Kilbride and I assumed the job would be there, near to my friends in Glasgow, but no it was in

Munich. 'Well, why not?' I thought.

I was blown away to find a welcoming committee at the airport. The department I would be working in had fourteen different nationalities and at least half of them were represented in the tavern that evening. It was clear they all had a good social life together and the evening session was to assess whether I would fit in. I thrived in the social whirl that I'd been missing in Aberdeen. My colleagues in Munich spent their money on leisure activities rather than possessions – ski-ing, wind-surfing, tennis, music and theatre. It certainly taught me that what you do in your life is more important than what you own.

After two years I transferred back to Scotland with Motorola. I was still at the work hard, play hard stage, and I enjoyed the demanding challenges thrown at us. A big difference I began to notice between the office culture in Munich and East Kilbride was that the Scots were much more demanding and much less forgiving. One of the worst things about the people management in East Kilbride was what they called 'Ranking and Rating' which we all referred to as 'Ranting and Raving'. We were told that this was a company-wide process, but I didn't experience it in Munich and I certainly never heard colleagues in other European factories refer to it, but it started to create an atmosphere of blame and toxic competition. In addition to an annual appraisal, every employee was assessed on how good they were at their job, and scored on their future potential. If you scored well you were encouraged to move abroad to get a better understanding of the worldwide corporation, but if not then your employment prospects instantly became bleak. During the six years I worked there, the bottom 10% were systematically 'let go' each year – shown the door just days before Christmas. I only ever lost one member of my team that way. He'd only joined the organisation a couple of months earlier, and the cruelty of what was done to him stuck with me forever.

In the mid-1980s, I knew I was ready for rural living again. Two friends and I bought a large house with two acres of land in a small village in Ayrshire. The life of my grandparents, a life I had wholeheartedly rejected as a teenager, suddenly became attractive again. Instead of a cow, hens and a dog, we had a horse, hens, two dogs and cats. To complete the menagerie, one day my uncle arrived in his tiny Honda Civic with a hive of bees in the back.

I decided it was time to move out of the corporate world and into the public sector. I naively thought that being an IT manager in an NGO would be straightforward; that everyone would be working together to get the best outcomes for Scotland, but it was an eye opener. The phrase I remember hearing over and over again was, 'we will take appropriate action' which basically meant that they'd considered an issue but would do nothing. It was beyond frustrating. I just couldn't understand the culture of complacency.

About a year after my move into the public sector, I was beginning to realise that I was the problem. By my mid 30s I had reached a stage in life where I just didn't fit in anymore. I wanted to get out of the rat race and do something real, something meaningful.

I attended a self-development course that started with envisioning where you wanted to be in five years' time. I wrote down that I wanted to be living in the country making something, not playing corporate games. When I realised we were about to spend the next two days working on a plan of how we were going to achieve our goals, I ripped up the piece of paper and wrote down that I wanted to be head of the IT department. I could pretend to the course trainer that I was on the right career path, but I could no longer pretend to myself. I'd had enough.

Chapter Three

David

As Dad approached 65, he was ready for a change and wanted to step back from farming. At the time my employer was undergoing re-structuring and very generous packages were being offered. I was ready to come home. Voluntary Premature Retirement paid for our move from Shetland to Galloway and allowed me to buy into the family business. At 32 I now had the perspective to appreciate the good things about farming, and having control over my own life, but coming home was tough. It was also clear by this point that my marriage was on the rocks and the end was rapidly becoming inevitable.

Little had changed on the farm over the intervening twenty years and all the labour-intensive activities that had driven me away were still here. Things needed shaking up, modernising and intensifying. We didn't even have a four-wheel drive tractor or a quad bike, both essentials on a modern farm. I hadn't come home to continue the same old humdrum routines I'd grown up with. I had plans! I'd seen a bit of the world, admittedly mostly a Scottish farming world, but I was aware that global changes were afoot. There had recently been an international trade deal done – the General Agreement on Trade and Tariffs – that would allow imports of cheap food with a potentially devastating impact on the UK's farming industry. The buzzwords of the moment were efficiency, productivity and diversification. I decided we would do all three.

More animals, more purchased feed, more fertilisers, more pesticides, more antibiotic, more drugs and drenches, more machinery, more technology. This, as I had preached to my farmer clients, was the future. Of course, there was resistance from the staff, no one likes change. There was Bob White, the foreman, who had been born on the farm. His first job, when he joined the farm team at age 14, was to dig the dead ewes out of the snowdrifts of the great snow of 1947. Bob wasn't tall but he was as strong as an ox and he had a knack for cobbling stuff together to keep things going, a very useful skill on a farm.

Then there was Jim and his younger brother Billy McKnight, both in their late twenties, married and starting families. Their dad had recently retired from the farm, and the boys had both worked here since leaving school at fifteen. They knew the farm and the farming ways much better than I did. Jim in particular had many skills and was perhaps the most resentful of my seemingly reckless race for change and disregard for their opinions.

Last, but not least, was Jock and Phemie McCarlie. They had been dairy-people and calf rearers for forty years, and cheesemakers until 1971. Jock had served in the forces in the Far East at the end of the Second World War. No one argued with Jock and, since he was due for retirement, luckily I didn't have to. I joked with him that at least in dairying there wasn't a 'Monday morning' – he got one weekend off a month and two weeks holiday a year. His reply was, 'Every morning is a Monday morning.' The guys were facing up to the new reality that things were going to change, but Jock had had enough. Farm life became full-on, and while I was making changes to the farm, I couldn't help but notice the subtle, and sometimes not so subtle, changes in my own family.

There was always a reason the kids couldn't come out and help on the farm and whenever I had some spare time, I found Jane had taken them off to see a friend. I suspected there was some kind of weaning process going on.

Farm work with the new, modern approach was all-consuming and, to be honest, it was an escape from the growing unpleasantness of family life. With the increase in animals on the farm came a disproportionate increase in health and mortality issues – diarrhoea in the calves, mastitis and lameness in the cows and sheep, ringworm, blindness and worm parasites in everything.

We spent a huge amount of time running around plastering over the cracks, but never really getting to the root cause of the problems. I was deeply uncomfortable with these outcomes but reassured myself that these were merely teething troubles and that it was just a matter of bedding-in the new system. After all, everyone who was progressing in farming was doing it, so it must be right. Right?

It was at this time that I first began to experience utter bleakness that I was unable to shake off. It was there first thing in the morning, and it followed me around like a personal black cloud until I got off to sleep at night. I had experienced short periods of anxiety in Shetland from the various stresses I'd encountered there, particularly in the early days, but this was different. It wasn't associated with anything in particular and it hung around for weeks. We have something of a history of mental illness in our family, but I wasn't prepared to concede that I might be suffering from some sort of depression. That, I concluded, would be a sign of weakness. Head down, battle on.

As separation looked like inevitable, I talked through our marital situation with the family solicitor. 'You do not want to go to court,' he said. 'Your best option is to allow the separation process to happen and allow the wounds to heal.' Sound advice. Fighting would only further damage my relationships with the children.

Depression was overtaken by a sense of utter despair. I sensed I'd already lost Mark, at thirteen years of age he blamed me for the separation, but I'd be devastated if I lost the girls as well. I managed to negotiate a couple of hours with Margaret and Christine each weekend to build a cross-country obstacle course for their ponies. Normally I had little time for ponies, but this was a small price to pay.

Eventually there were signs that a move was imminent. I had to say something to the girls. I had to explain things and tell them I loved them and would always be there for them. By chance an opportunity arose to take them aside with no one else around. We sat holding hands, but all I could manage was, 'I'm so, so sorry...', then tears and hugs and I was gone.

A day or so later, they left overnight. A midnight flit. I had no idea where they'd gone. I didn't see them again for three long, excruciatingly painful months. This grief, of being forcibly separated from my children, haunted me. I became something of a recluse and buried myself in farm work. There is, after all, always plenty to do on a farm.

Most days I'd start at 6am and jog round the farm feeding the various groups of cattle and sheep to arrive back for the team start at 7.30am. When I say 'feeding' I'd give them a cereal and soya-based feed and then, if in winter, fork out silage, which is grass that's been fermented for winter feeding. It certainly kept me fit.

If I was on dairy duty, I'd be starting at 5am and would see to other stock work after milking. I'd have a cup of coffee and join the farm team with whatever jobs they had on for the day. It might be erecting fencing, sorting dry-stone dykes (walls) – there are many miles of these on the farm – or it could be repairing blocked field drains, preparing fields for sowing oats, barley, grass, turnips and kale.

Then there was the various sheep and cattle work – sorting sore feet, treating against various ailments with vaccination, drenching and dipping. There was lambing, shearing, drawing lambs and cattle for market, and so on. The land needed sorting too, spreading fertilisers, spraying weeds, making silage, making hay, hoeing turnips, shawing turnips, moving temporary electric fences, sorting old machinery, logging windfall trees for firewood and, if we could find the time, building new farm buildings to house equipment and cattle. It's easy to lose yourself completely in the daily demands of a farm.

Friends were kind and would invite me for supper now and again, but I was poor company. Sometimes there would be a single young woman at the table, usually sat next to me, but I wasn't interested. I was a mere shell. My self-confidence was absolutely shot-through.

When the children did occasionally come for a holiday, it was like a shot of adrenaline. They helped on the farm, mostly with the sheep, the way I'd always dreamed they might. When they went back, they left a hole in my life and the black cloud returned.

A year went by and then two. I couldn't go on like this. I realised I needed a soulmate or things would end shortly and badly.

Living and farming in a small, fairly isolated rural community wasn't the ideal environment to meet such a person. To all intents and purposes the internet, as we know it today, didn't exist but I joined a computer-based dating agency called Dateline and met a couple of young women through that. Not really my types. 'So much for computer dating,' I thought, 'this isn't going anywhere.' There was a small window of free time between our first cut of silage and hay making, so I gave it one last chance.

I wouldn't say I was entirely blown away by my first impressions of the city-girl look of Wilma – frilly blouse and flowery trousers. We began to chat and walk and more chat. I found a connection with her intellect and energy. This might be worth a follow-up, I thought, so we agreed a date after the hay making but before the second cut of silage.

For the first time in many years, I had something positive to look forward to. I think it was our third date when I popped the question, 'Would you like to see my business plan?'

Chapter Four

Wilma

Nothing has been conventional about David and me, not even how we met. It was in 1992, David and his wife had separated two years earlier and my own relationship had broken up at around the same time. I was working in Glasgow and he was on his farm 100 miles away. There was a lot of stigma about computer dating then, to the extent that I never told my parents how we met. I reckoned they would think it was bad enough for me to be with a man who'd been married before, but if he was reduced to finding a partner through computer dating, then there had to be something wrong with him, and I guess by implication, something wrong with me.

We met for lunch at a remote hotel halfway between our homes. The main things I remember were his unfashionable white squash shoes and the hole in his jumper. On this alone I dismissed him as a prospective partner, and perhaps that's why I relaxed and started to tease him a little. There was lots of laughter and lunch went on for much longer than either of us expected.

That evening a friend asked me how my date had been. 'Not a hope, you should have seen what he was wearing' I said, 'though he does have a good sense of humour'.

'Maybe that's more important than fashion?' she counselled, and here we are, thirty years later.

That's not to say I didn't have some doubts in the early days. David was definitely still hurting from the breakdown of his marriage. I had only seen divorce through the experience of a couple of female friends who'd been abandoned by their errant husbands, so I did wonder what on earth David could have done that had made things so very bitter. I told him of my concerns and he just shrugged and said he'd have been surprised if I hadn't been wary.

A couple of months after David and I met, I noticed a lump in my breast. The doctor referred me to a specialist who said it would just be a cyst and he gave me the choice of having it removed or not. I definitely wanted it removed but there was no urgency, so three months later I had an overnight stay in hospital to remove the cyst. A week later I had an extended lunch break to get my stitches out when the surgeon told me it was malignant, and said he'd arrange for me to have a mastectomy the following week.

His manner was so unsympathetic – so cold and brutal – that I had to be helped out of the room by a nurse. She took me to a separate room and said she'd get me a second opinion. We went to the surgical ward and she introduced me to a younger surgeon. He looked at the report of the surgery already performed and concluded that the tumour had been removed completely, and said he didn't think a mastectomy at this stage was essential. Instead, he proposed another operation to check my lymph nodes to see whether the cancer had spread and would then decide what further treatment was necessary.

The uncertainty of the next couple of weeks was just awful. I felt cheated. My mother had a mastectomy when she was 57, fifteen years previously when I was still a student. I was only in my mid-thirties, my life hadn't really got going yet. All the things that I had been anticipating in my new relationship with David could be cut short. And could I actually put someone that I had only known for five months through all this uncertainty?

Chapter Five

David

I'm not sure Wilma knew what she'd taken on. I was damaged goods. Huge insecurities, self-doubts, sometimes exhibiting explosive behaviour – and could I even be sure that Wilma was the right person to partner me on this next stage? Goodness knows what she thought.

Then she was diagnosed with breast cancer. In those days the outlook was much more uncertain than it is now. I couldn't cut and run to leave her to face this life-threatening crisis alone. She was put on a course of chemotherapy to be followed by radiotherapy.

During this period, before treatment started, we made a decision to take a few days away in which we visited my slightly eccentric Aunt May near Ipswich. I had the greatest respect for my aunt and the relaxing break was just what was needed for both of us.

I was slightly surprised on the morning of our departure when Aunt May took me firmly by the elbow and led me aside. She looked me earnestly in the eyes and said, 'David James, this girl, believe me, this girl is the right girl for you.'

Aunt May was right.

Chapter Six

Wilma

David wanted to come with me to get the results of whether the cancer had spread, but a farmer's life is never predictable. The appointment with the consultant was in Glasgow, a two hour drive from the farm. David did his early morning check of the animals but he didn't reappear at the time promised. No-one had mobile phones in those days, and I found him in a cattle shed helping the farm team with some repairs.

It's never a good idea to have a screaming fit in front of others, but I'm afraid I let myself down that morning, and the breakneck trip to Glasgow didn't help me find a sense of calm. To raise the tension even further it turned out that my results hadn't even come through yet, so my appointment was delayed by four hours. We went to my house to find a huge bunch of flowers from my workmates on the doorstep. To me it looked like a wreath and I've had an aversion to bouquets of flowers ever since.

Thankfully all the lymph nodes were clear. The oncologist proposed six chemotherapy treatments followed by four weeks of radiotherapy. I was still working in Glasgow, so I'd get the chemo treatment every third Wednesday, then drive down to the farm for a long weekend and be back to work on the Monday. I was a little more tired than normal, but it was only the final chemo treatment that caused any sickness. Other than one morning every three weeks when I sat in the treatment room, I could, to a large extent, forget about cancer.

Coming to the farm every weekend was not solely about me wanting to see the man I was falling in love with. The farm itself was beginning to feel like home. In fact there's a hill on the A75, just half a mile before the turn off for Rainton, that told me I was home the very first time I drove down it. The surrounding hills and the sea ahead seemed to put their arms around me and welcome me in. I give a sigh of relaxation every time I get to that point on the drive home.

David obviously wanted to impress me, not just with his farm but also with the place that could become my home. We would cycle to the beach at nearby Carrick to have a picnic, and I still tease him asking where the beach is. Coming from the north east of Scotland, the sandy beaches of my childhood home stretched for miles. Here you need to check the tide times to find sand.

I had two dogs who were even more excited than I was when I loaded the car on a Friday afternoon. No longer the morning and evening walk around a city park for them, they had acres to explore! We followed David around as he kept to his daily farming routine. The views around the farm are spectacular – swathes of green, hills, sea and islands all in one view. It felt like I was on holiday every weekend.

From the outside looking in, farming seemed like a lifestyle that would suit me. While much of it is a fixed routine, there are all kinds of interesting emergencies that appear from nowhere and need to be sorted there and then – even if the fix can be done with a piece of binder twine, the challenge of it appealed. These were tasks that mattered. We've had quite a few visitors to the farm who have said they envy us the fact we have to get out of bed every day; that we don't have a choice. Animals need to be checked, machinery has to be repaired, fences have to be fixed. There is so much to do that simply can't be put off, so I guess my workaholic-self was seeing farming as something more vital, and much more meaningful, than computer programming.

Throughout the period of my treatment David started to share his ambitions with me. He wanted to diversify the farm and wondered if I would be interested in running a business on the farm? This coincided with a re-structuring in the organisation I worked for and they were looking for volunteers for redundancy.

I clearly remember David calling my bluff. 'You can carry on working there making a comfortable living in a job you aren't enjoying, or you can take the money and do something here that you could shape'. It was a no brainer.

The idea behind diversifying the farm was to build resilience into the business. That would make the farm much less dependent on the wholesale price of milk, a price that is determined by the global commodity market rather than by the local cost of production. We would take the milk produced by the dairy and do something with it on the farm to add value to it. The farm produced a lot of milk – about 700,000 litres a year – so there was plenty of potential to create something new, exciting and ambitious.

The business plan for the new venture began with a trip to Cornwall to visit a cheesemaker, an ice-cream maker and a yogurt maker. Each of these products could be made with the milk from the farm, and we noticed that every one of these businesses also had either a café or a visitor centre. But that was Cornwall, a holidaying hot-spot. I remember on the long drive home, we agreed that ice-cream was likely to be successful for us; good quality ice-cream, made from simple ingredients. It was a product we felt we could make well, we figured it would be popular, and we knew there wasn't much competition. But David was clear on one thing. 'We are never going to get involved in tourism,' he said. 'There is no way

I want people walking round the farm'.

Around this time I raised the subject of whether we would have children or not. I have never been particularly maternal. I had female friends and relations for whom having children was more important than having a partner. I was the opposite, although I had noticed in the past that when I was in a long term relationship I started to think I might want children. And so it was with David.

Early on I knew that I wanted a mini-David. Of course there is no guarantee I'd have gotten a mini version of David, but that's what my body was telling my brain. David already had three children from his first marriage and said that if I wanted children it was fine with him, but I would need to take into consideration the demands of both the farm and any future business we set up. As it transpired, my periods stopped after my first chemo session and never returned. Initially I was very disappointed. It would have been ok if I had made the decision not to have children, but to have the decision made for me was a blow. That feeling didn't last long. Cream o' Galloway, our ice-cream venture, became my baby, growing into an adult that has never left home.

Chapter Seven

David

I had thrown myself into the farm intensification process, but it was achieving little. True, we were turning over more money than ever before, but there was little or no more profit. We were running faster just to stand still and I needed a new direction. Embarking on a farm diversification enterprise with Wilma to launch Cream o' Galloway ice-cream was exciting.

This was the early 1990s, the final decade of the last millennium, and we had our eyes firmly on the future. We were going to convert a semi-derelict 17th century farm building and we needed planning permission, building warrants and capital costings. At that time European money was available for farm diversification and the application process was surprisingly straightforward; a budget, a single A4-sized piece of paper and an interview. How things have changed!

With funds in place, we went in search of equipment to make ice-cream. There were very few farm-based ice-cream makers at that time and no off-the-shelf systems to buy and plug in. It was all going to be cobbled together with bits from here and pieces from there, mostly second hand. Our ice-cream freezer stores were refrigerated shipping containers.

That summer we began work preparing the building – clearing out the accumulated junk, digging out the ancient sandstone flag and hand-hewn, granite cobblestone floors, picking out the lime mortar from the old stone walls and re-pointing them, taking off the old Lancashire-slate roof and ancient timbers.

We were stripping the old building naked – its ancient purposes exposed, then swiftly gone forever. The opposing doors that would allow the wind to blow through and remove the light chaff produced by the winter-long, hand-threshing of the oats; the slit windows that ventilated the oat sheaves stacked inside; the farm workshop and cart-shed with bothy room above for the itinerant seasonal workers – all of it had to go, along with the cobble floored stables for up to four draft horses and the byre for ten cows. All gone, making room for progress.

I don't know how many lives I've been allocated. I guess we'll find out in due course, but I've certainly used up quite a few. That summer I used up three of them in a single month. First up was the ice-cream waste-water storage tank. We were constructing the tank ourselves. Bob was on the JCB digger, and the hole was about ten feet square, and ten feet deep. Young John Porter, who had recently joined the team straight from school, and I climbed down a ladder to check the dimensions at the base of the hole. The soil was stable, compacted clay and looked pretty safe – it would only take a few seconds after all. We checked and it was good. John was halfway back up the ladder while I was standing at the bottom, holding it against slippage. Then, something hit me on the back of the leg throwing me against the wall. I looked back to discover a three-tonne slab of clay had broken off the far wall and lay where I had stood a few seconds earlier. It had happened so quickly that Bob hadn't even had time to call out a warning.

The next incident happened a week or so later when I was bringing home a heavy load of railway sleepers which were to be used to cover the waste-water storage tank. Normal railway sleepers are pretty heavy things, and one person can struggle to lift them.

These were extra-large ones, about half as heavy again.

I was towing the single-axle trailer with a light Subaru pick-up, and I was taking my time, going at a sensible 30 or 40 miles an hour, when the trailer began to 'fishtail'. There was nothing I could do. The fishtailing became more and more violent until, within a few seconds, the pick-up and trailer jack-knifed and spun a complete 180 degrees, spewing the load of sleepers across the A75 – a busy, fast main road traversed by hundreds of lorries every day, thundering along at speed on their way to and from Northern Ireland.

Thankfully the pick-up didn't roll. I'd come to a halt on the verge. I clambered out and surveyed the railway sleepers with absolute horror. This was just over the brow of a hill and at any moment an Irish truck and following traffic was certain to appear and plough into the scattered sleepers. By great good fortune a forestry worker had been driving up the other side of the road. He stopped to help and together we dragged the sleepers clear just in the nick of time as a column of lorries and cars swept over the brow of the hill. His actions undoubtedly saved lives that day.

It's funny how disasters, or near-disasters, tend to arrive in threes. There had been a leak in the three-storey, farmhouse roof and I'd been deferring the job for several weeks because we'd been so busy, and besides, I'm no good with heights anyway. An opportunity arose to do something about it, so I found some sealant and the applicator, a spare slate, the extra-long set of ladders and the ladder with the hook-over end that we used for roof work. I went up the long double ladders and got the hook-over roof-ladder into position, hooked securely over the ridge. Uneasily I crawled up the roof-ladder, found where the water was getting in through a broken slate near the ridge and fitted the new slate with the sealant. I had been at full stretch across the roof from the ladder, putting a lot of strain on the hook-over end which flipped up as I tried to pull myself back onto the ladder. The ladder and I slithered at speed down the steep roof before my fall was broken, by sheer luck, as I crashed through the top of a little coal shed to the rear of the farmhouse, landing heavily on the bags of coal, bruised and shaken, but thankfully nothing broken.

By early August we were ready for the builders to start work on the ice-cream factory, as contracted. The weather was beautiful, warm, dry and settled right through to late October, when the builders finally got here – seven weeks late. Then, on their first week, the skies opened and the rain started, temperatures fell and by early November the rain had turned to sleet then driven snow. My main memory is of the builders working on scaffolding with hats pulled down, heavy coats flapping in the bitter wind, fingerless gloves, and scarves wrapped round their lower faces leaving just a slit to see through. I didn't envy them but thought, heartlessly, if they'd been here when contracted they could have had the roof up by now and be working inside. We were already two months behind schedule.

We had planned the launch of the ice-cream at the Royal Highland Show in 1994, a major event that takes place just outside Edinburgh each June. In those days the show was somewhere you could make contact with buyers from the retail and catering industries, an ideal shop window for small, rural, start-up food businesses.

Although we had made up some of the deadline slippage, we were still finishing off the building work as we started ice-cream making. Painting floors, concreting the yard out front, installing fan ventilation and indoor cold stores. I can remember carrying boxes of ice-cream tubs along walking-boards over wet cement to the van heading for the show, stress levels through the roof. By great good fortune, my three children Mark, Margaret and Christine were here – now aged 18, 16 and 14 – and they were an enormous help, we couldn't have done it without them.

The ice-cream was in a shiny new van driven by Wilma, Mark towed a borrowed caravan with a borrowed Land Rover, I followed with the girls in the beat-up pick-up, towing a home-made trailer with all the show paraphernalia. We set up the stand in preparation for the following day and after a well-earned meal, we headed towards the caravan in the nearby park, exhausted and ready for a good night's sleep.

I must say I was apprehensive about the caravan site. Talking to a seasoned campaigner some days before, on hearing that we would be staying in the caravan park he said, with a dark look, 'Ah yes. Most folk do that. The first year.' We were all so tired we were bound to sleep, but the night-time entertainment started about 1am with the inebriated young farmers returning. Things quietened down by about two, but then trucks started roaring into the neighbouring chicken processing factory at 4am. That eased off by five but at 6am the planes from neighbouring Edinburgh Airport started up. Ah yes, most folk do that, the first year... I could see the young farmers weren't so daft. An alcohol anaesthetic was probably very useful here.

The stand in the food hall was ready for the gates to be opened and as we stood there someone opened a small tub to check quality. 'Ugh! This tastes of diesel!!' A sense of dread rose. We were about to launch with ice-cream that had somehow got contaminated with diesel. How on earth had that happened? Quickly we opened more tubs, more flavours. It was just the banoffee. Thank goodness, but how had it happened? Then it hit me. We had used a diesel air heater to speed up the drying of the floor paint. At one point the heater had cut out and had pumped out diesel fumes into the ice-cream dairy. It was a miracle only the banoffee was affected. We quickly replaced the banoffee tubs with vanilla, and the crisis was averted.

We had hired a time slot in a small marquee for the official launch to which we had invited members of the press and buyers for the retailers. About twenty folk turned up, possibly some were bemused passers-by. Wilma and I did our talks, and the wow-event was a 'champagne' cascade with the glasses containing a small amount of ice-cream which looked quite impressive. The whole thing seemed to have gone well, with assurances of media coverage and future sales opportunities to be followed up.

Returning to our stand we started an adrenalin-driven four days of samples, tastings and sales – in fact it was going so well Mark had to drive to a meeting point half-way home to replenish stocks at the end of the second day. The launch was a great success. We were on the road to being a player in the Scottish, and perhaps the UK, ice-cream market. We were a wee bit unlucky in that Mackies had beaten us to the market with a decent quality, but less expensive, farm-made ice-cream supported by a substantial marketing budget funded from the sale of their milk retail rounds in Aberdeenshire. Nevertheless, Wilma and I whizzed around the country clocking up huge mileages doing tastings and sales pitches to whoever might listen. Occasionally we got lucky.

As an after-thought to the ice-cream dairy, we'd added a small ice-cream parlour and teashop, and a playground, which we advertised as being open from the first of July. As ever, we were running late and we were still finishing off building the playground and covered areas the night before opening. Everything was home-made. We'd visited some local play areas and measured up see-saws, climbing frames and such. If it was good enough for them, it was good enough for us.

One thing that was different – very different – was the straw-play barn. We'd never seen one at any play park we'd visited. Soft-play, yes, but straw-play, nope. We knew from personal experience that kids love straw-play. We created dens and tunnels with the bales in what had been the old 'tractor shed'. Just to be sure it was safe, we tied the bales together to avoid cave-ins. No question, kids loved it! Adults hated it because the kids would strip off a jersey or shirt in the back of the car on the way home, spilling out handfuls of straw, but still, they came back. Our only concern with the straw-play was occasionally finding a member of the public supervising the kids with a lit cigarette in their hand...

Like the ice-cream, the visitor side was a bit slow the first year. This was in the days before the internet and word travelled slowly. Then when, at vast expense, we got signs on the main road and leaflets in all the local accommodation providers, things began moving. Ice-cream and adventure playground was a recipe for success that we'd accidently stumbled across. Visitor numbers began to rise sharply, and we were forced to play catch-up as our meagre offering was overwhelmed. We were adding activities for the visitors every year, but it was never quite enough.

We gradually began to recognise that we had a little gold mine on our hands, so we introduced a series of major investments in car parking, outdoor play activities and a major expansion in wet weather facilities. Ice-cream manufacture was barely washing its face but the visitor centre – as it was now grandly referred to – was paying the bills, and more.

Chapter Eight

Wilma

The early days of Cream o' Galloway are now a frantic blur. We were young and naïve and we'd attend any event, no matter how small, in order to raise our profile. That initial summer was busy and we felt we just couldn't fail, but then winter came, the tourists left the region and we had virtually no customers. When we first mooted the idea of making ice-cream to David's family, his father kept talking about there not being enough chimneys. The reality dawned. Once the tourists left at the end of the summer, our region just retreated into itself. We originally thought we'd only need to sell within Dumfries & Galloway. It's a huge area – 100 miles from east to west – but it's very sparsely populated. We needed to take our ice-cream to where most people live in Scotland – the central belt.

Our business grew slowly. Even when we got a listing in supermarkets in Scotland it was only a gradual increase in sales, but at least it was going in the right direction. We were optimistic we could eventually be profitable, but we wondered whether there was something more we could do.

From the very first time I'd visited the farm, I was keen to convert it to organic production. I had no idea what that meant practically, but having read numerous articles in the Guardian, I felt it was something to aspire to. Having said that, for it to happen it needed to be something that David wanted too. He was open minded when I talked about converting to organic farming, but didn't know how to make that transition. For both of us our motivations were a mix of wanting to have less of an impact on the environment, but to also tap into what was, by then, rapid growth of the organic market.

David wasn't sure how his father would react to the idea of converting the farm to organic production, after all, he'd been extremely worried about us diversifying the farm a few years earlier. During the 40 years he had managed the farm, David's father had made major agricultural improvements. Letting it 'go back to nature' – in other words, be covered in weeds – could have been taken as a slap in the face and throwing away his life's work. On the other hand, his father was becoming concerned at just how much money was being spent on vet bills and farm chemicals. David's mother was very supportive but his father was a bit more circumspect.

David visited many organic dairy farms throughout the UK to get tips on how it might work for our farm. I quickly realised these business trips were as close as I was going to get to a holiday, so I would join him when I could, but still he was holding back. This was mainly because the Agricultural Department in Scotland had announced they would soon introduce a new support system for conversion to organic farming in Scotland. No point in jumping too soon and missing additional support. I was becoming frustrated. It was clear the organic market was booming and I worried we were missing the boat. At that point there was only one company in the whole of the UK making organic ice-cream, I felt we needed to launch an organic product range quickly before someone else did.

David and I married in January 1999. We had a small wedding with just sixteen people there – well really it was fifteen people, with David appearing at the critical moments. He was unwell – 'must be the stress' – and lay in the back of the car throughout most of the reception.

The initial plan for our honeymoon was a mini tour of the UK visiting relatives who hadn't been able to attend our wedding. That included my parents, and David's much loved Auntie May. David definitely wasn't up for the constant food and drink that would entail, so, we delayed the honeymoon by a couple of weeks and headed off to Paris instead. I'd bought a newspaper for the journey and noticed an article announcing that Mackies were converting their farm and their ice-cream range to organic. Mackies were and still are the biggest independent ice-cream makers in Scotland. We have always had a lot of respect for them, but there was no way they were going to launch an organic ice-cream range before we did, so the acceleration of our organic conversion plans began on our honeymoon!

The major change in becoming organic was, of course, at the farming level. There is a minimum of a two year conversion period before farm produce can be certified as organic, which effectively meant we couldn't use the farm's own milk in our organic ice-cream range until 2001, so we needed to source organic milk from elsewhere. Fortunately, a family friend, Graham Keating, was the MD of Yeo Valley Organic at that time. Skimmed milk powder is one of the ingredients used in the making of yogurt, so David travelled to the other end of the country to pick up half a tonne of organic milk powder that we could use to make ice-cream until our own fresh milk was certified. Following frantic sourcing of all the ingredients required for an organic range, and the design and printing of packaging, Galloway Organic ice-cream was launched in the summer of 1999.

We learnt lots of lessons over the next couple of years, and the main lesson was that the market for organic food is mainly in the south of England. It still is. We had genuinely believed we would be phasing out our entire non-organic ice-cream and replacing all the flavours with 100% organic alternatives. At this stage our non-organic range had listings in three supermarkets in Scotland and so we enthusiastically replaced all those products with the new organic versions. All the supermarkets came back to us within just a few months to say sales had slumped and if we wanted to keep our place in their freezers we should revert to the original, non-organic range immediately. Meanwhile, a week visiting delis and cafés in London resulted in opening 40 new accounts down south. The key interest for those new accounts? That the range was organic.

Why should there be such a difference? Obviously disposable income is an important factor, but so is the perception of organic food. In a heavily populated area people are naturally concerned about pollution and the effect it has on their health and wellbeing. In London buying organic food is a positive step for your health. However in Scotland, even in the heart of our biggest city, you're never far away from green fields, hills and lochs. Perhaps it feels as though everything produced here is as good as organic without the need for the label?

Scotland's hesitance over organic wasn't going to deter us. After all there were more people living in London than in the whole of Scotland, so we could have a product range for Scotland and an organic range in England. The positivity in London even extended to the supermarkets who were keen to launch own label organic versions of their ice cream. This was beginning to feel like a step that could lead to meteoric growth, and we didn't have a factory that could cope.

We were beginning to pay off our initial loans for the original ice-cream factory, so now it was time to expand further. I'm much less of a risk taker than David and I questioned whether it was sensible to be taking out bigger loans to satisfy a market that wasn't on our doorstep. On the other hand, I understood we were finally seeing a real opportunity for business growth – we had to be ready to meet it.

Eighteen months of more business plans, architect's plans, securing funding and construction. We had an extension ready by the summer of 2000 and filled it with the luxury of reliable equipment. We were on a roll.

Chapter Nine

David

Within the farming community, the premium price for organic produce was legendary. The Scottish Government had finally relented and was offering to help farmers convert to organic. We had been visiting and cross-examining organic farmers in England for a couple of years to learn as much as we could from them. The time was ripe. The stars aligned. We didn't have much of a clue about what we were doing, but at least we had guidance from those who had gone before us. As with any major change, the most difficult part is the change of mindset.

We were moving from a technology-based farming approach to one that would be management-based, but which used technology to help us achieve the desired outcomes. A common mistake is to look for organic interventions to replace those conventional interventions we had become dependent on and addicted to. What were the organic equivalents of fertilisers and crop stimulants? What organic pesticides could we use? It turned out that way of thinking was just wrong headed. We learned to step right back, look at our farming system and figure out how to avoid needing to use those interventions in the first place. That is a big but crucial step in mindset change and it ain't easy.

The next problem with nature-based solutions is that they take time to feed through – a lot of time. We had been used to seeing results pretty well instantly with conventional measures, but nature responds much more slowly. More often than not it will take many years – and for us it was about ten years – before the soil and crops recover from their chemical dependency, created by heavy applications of fertilisers and pesticides over many years. Of course, during those ten years we were getting better at understanding the needs of this approach, and managing and facilitating those needs. As we learned, we started to see grass yields and stock productivity return to the levels we had previously been achieving with those chemical crutches.

In the circles we were now moving in, climate change, biodiversity loss and pollution were all major issues. Organic farming was a big step in the right direction, but we wanted to go much further, into renewable energy, plastic recycling, stopping any pollution from the farm and, best of all, tree planting. Not just the usual conifer planting, oh no, we would plant mixed, broadleaf trees, they were best for carbon sequestration and biodiversity. But there was a slight problem. We knew nothing at all about tree planting, we'd only ever cut the things down.

I made some enquiries about organic tree planting, but it turned out no-one planted trees organically. You plant them conventionally with fertilisers and pesticides then, after a couple of years, convert the woodland to organic. I was deeply uncomfortable with this. Not just because I felt it was a bit of a fraud, but because I knew the local rural rumour-mill would go into overdrive – 'The Finlays are cheating! They're supposed to be organic and getting all those organic subsidies, but they've been buying fertilisers and weedkiller!' I could just imagine it. We'd never be able to shake off doubt about our honesty, so we decided to try it organically – no fertilisers, no weedkillers and no insecticides. By good fortune we noticed an advert by an 'ecological forestry' consultant, Alistair

Bartholomew. Ecological? That sounded just the job.

We started our organic tree planting experiment on a small five-acre plot of poorish land near the visitor centre. This was going to be the visitors' dog-walking area. Alistair was sourcing all the materials for us – trees, vole guards, rabbit wire – and putting together our grant application. We installed the rabbit fencing and 'scare wires' to stop deer getting in. Alistair arrived, demonstrated how to plant a tree and explained which species were best for each spot – alder and willow in the wet areas, ash and oak on the relatively deeper soils. Birch, hazel, hawthorn and Scots pine went in the drier parts, with holly and wych elm scattered here and there. The mix of species was supposed to reflect those that would have occurred here naturally.

There were about 2,000 of the delicate wee twigs and it was hard to believe they would ever become full grown trees, reaching 30, 40 or 50 feet up into the sky. One of the biggest threats to young trees is that grasses and other plants can smother them. We had purchased a large quantity of hemp mats, half a metre square with a thin latex covering on one side. They were a bit pricey but they looked just the job; a mat fitted round the base of each tree which would keep the competing weeds at bay.

We set to work and over the next few weeks and, with the help of members of the family who were keen tree huggers, we got the trees all planted with their vole guards and mats in place. Job done. But while we had stymied the attempts by the voles to eat the bark off the young trees with the tree guards, they retaliated by chewing up the mats and using them, presumably, as bedding. Within a few weeks of planting being completed, the hemp mats had completely disappeared. We had to go back in and use straw as mulch around the base of the vole guards. It was a lot of work but it worked.

That was our first venture into the world of tree planting. Over the years those trees thrived and grew rapidly. It felt good. My father, however, didn't agree. Dad had run the show here, his way, for 40 years with no interruption from sons or daughters. Everyone knew him and he was quite popular. He had hoped to 'retire' into politics, certainly as an activist, leafletting, organising meetings, fundraising, recruiting and so on. Perhaps he even had hopes to stand for election in some form or other, but he was no 'yes man' which effectively ruled him out as a party candidate.

We had never really got on and his 'tough love' approach didn't help. He considered me lazy and impudent, which was probably true. Father and son relationships are often quite difficult, and I've never been good at managing the emotional stuff. While I frequently want to express my love for him, there's just so much baggage that gets in the way.

I guess things started off equitably enough after I took over the farm. He was sufficiently distracted by his new interests to give me some breathing space as I cranked up the fertiliser and feed rates, increased stock numbers and got some decent machinery. That all seemed to go down well enough with him, just so long as I didn't go into debt.

Our relationship started to deteriorate when we were planning the Cream o' Galloway diversification. I had crossed the 'no debt' red line, and crossed it in a big way, and it really bothered him. Dad claimed he was going to move out of the farmhouse once we opened the farm to the public, but once Cream o' Galloway launched he found fertile new ground for his conversations and stories. We'd find him there at coffee and lunch times chatting to customers. He was in his element.

I decided to create a network of paths around some of the farm fields and in one we blocked up a drain into a bit of ground that was always flooding. We decided this would create a nice little wildlife pond for both nature and visitors to enjoy. That spring the pond was full to the brim and looking really lovely. It was a busy lambing time and, as usual, people got tired, tempers were short and nasty exchanges made. I got a phone call from Wilma reporting that the new pond was empty.

I zoomed down to the pond on the quad bike and examined the site. Sure enough the pond was empty and there were signs that someone had cleared the drain that I had carefully blocked up. I headed straight back to the farm, found Bob and asked him to take the JCB digger down to the pond, dig out the drain and fill the drain track with soil. This little event was referred to as 'the battle of the pond' and it was an indicator of things to come.

Talk of going organic didn't seem to faze Dad. He had been concerned by the rising fertiliser, pesticide and vet bills anyway so, on the whole, he saw transitioning to organic as a good move. He was a little more ambivalent about tree planting, but as our first woodland planting was on a small piece of poor quality land, he didn't object.

That all changed when we began planning the second phase of woodland planting. This time it was 50 acres of permanent pasture that was at stake, good grazing ground, and our intention was to cover it in trees. Before we could complete the application for approval, our ecological forester consultant contacted me and said he'd have to withdraw. Baffled, I asked him to explain. He said, 'I've had a letter from your father threatening legal action if I continue with this application.'

That triggered a spell of open hostility where Mum had to act as the go-between. It got so bad at one point that I remember Dad taking a swing at me. He missed. I grabbed his bike and threw it over the wall into the field. Things really couldn't get much worse.

It was at this time that a group of German school children were visiting the area on a two-week exchange with the local school. The school was looking for a host for one of the children, Maxi, and his mother. They needed somewhere with downstairs sleeping accommodation because Maxi used a wheelchair, so my folks volunteered. Dad went to a lot of bother adapting the farmhouse entrance for wheelchair access and even practiced driving around on the quad motorbike which, out of principle, he had never ridden before. His plan was to take Maxi round the farm on his wheelchair in a small stock trailer pulled by the quad. Against my expectations, this all worked remarkably well, and the visit was deemed a great success. It least it was a distraction for Dad.

I was on milking duty the weekend after Maxi and his mum had left and was deeply concerned when Mum arrived in the milking parlour at 5.30am in floods of tears. 'What's wrong?' I asked, dreading the worst. 'It's your Dad! He woke me up and said, tell David that the visit from Maxi has made me realise there is more to life than fighting over some trees. Go ahead with the woodland, but dedicate it to Maxi.' After Maxi, Dad backed off quite a bit but still managed to challenge some of my more rash ideas, which wasn't a bad thing so long as we remembered to keep it measured.

I had never really stopped for long to think what the impact of tourism, ice-cream, woodlands, ponds, debt, and the rest would be having on someone for whom much of this went against everything he believed in. Perhaps if I'd been more sensitive to his more cautious ways, we'd never have done half the things we have. Would that have been better or worse? I don't know. Certainly, the journey would have been less stressful, but maybe we wouldn't be where we are now. In the end we planted a total of 35,000 mixed broadleaf trees, including a nature trail dedicated to Maxi. The woodlands provide pleasant walks for visitors, a biodiversity boost to the farm and future crops of timber.

We frequently get asked on social media, often by vegan activists, why we don't plant oats and make oat milk, as if producing milk, or a milk-like substance, was our goal here. Of course, that's not our goal. On our farm, in this part of the world, on this hard, stony land, there are only two things we can grow well – trees and grass. We grow both, and they do different things. The grass produces high quality nourishment by way of grazing livestock, and the trees provide habitat. They are both valuable activities and they complement each other well. We started planting these woodlands more than twenty years ago. Now and again I'm struck by the maturity of these trees, towering over me, reminding me of the length of the journey we've been on.

As we headed towards the new millennium it was time for a step-change in our clunky, labour-intensive manufacturing processes and a major upgrade in storage and office facilities. We knew we would be stretched financially but with the market growth in organic we were confident we'd be ok, we'd done the budgets. On top of this production expansion we also upgraded our tea-room, toilets and car park to accommodate the expanding numbers of visitors. It was an exciting time! This work was all done by our farm team in the quieter winter period led by an exceptional individual, Jim McKnight.

As a child Jim hadn't been academic and so he'd left school as soon as he could, but he had incredible skills as a master builder. His skills and confidence started in small ways but by the end he was project managing the practical side of building a £1.5m farm dairy complex. He, and anyone available to help him, also built all of the adventure playground, barn, toilets, tea-room, nature trails and car park at the visitor centre. We owe much of what we've achieved here to Jim's exceptional skills.

Chapter Ten

Wilma

Looking back, Foot and Mouth Disease (FMD) in 2001 seems like a dress rehearsal for Covid-19 twenty years later, except while FMD only affected rural communities in the UK, Covid has affected everyone. For us, Covid felt easier emotionally than FMD for that reason alone.

Back then it was painful and bewildering watching the rest of the UK going about their normal day-to-day lives while we were in lockdown, confined to our farms. As a society we talk a lot more about mental health than we did twenty years ago. Now it's 'OK not to be OK'. It wasn't back then. We were not okay, and we know farming families who are still affected by that time, traumatised by the loss of prized herds a generation ago.

Whenever there's a survey on FMD, there's always a standard question – 'Were you affected by FMD?' Too right we were. The supplementary question is then, 'What animals were culled?' Well actually, none of our animals were culled. We may not have been infected, but we were certainly affected – emotionally, financially, physically and mentally.

We likened it to a war zone. 2001 was an election year and the government just wanted FMD to end quickly, no matter what it took. It seemed every decision was taken on black and white criteria without much common sense being applied, there were 'just get it done' contracts on culling farm animals. Some farms actually benefited financially from being culled out, and rumours were rife about farmers who had deliberately brought it onto their farms. Some neighbours don't speak to each other to this day for that reason.

Farmers who thought they would simply be taking their turn in Chairing their local NFU branch for a couple of years found themselves as high profile negotiators on behalf of a destroyed industry, and as counsellors for friends and neighbours who had lost everything. There is a direct comparison with the publicans and hoteliers who have found themselves in similar circumstances throughout Covid. A voluntary role they thought might take half a day a week suddenly became a 24/7 commitment from which they may never recover.

For us FMD was both financially and emotionally crushing. There was a permanent knot of fear in our stomachs dreading the daily announcements of how close FMD positive cases were to our farm. If it came within three kilometres of the boundary of the farm, our animals would have been culled. If any person or vehicle came on to our farm, who had previously been on a farm which had subsequently been found to have had FMD, then our animals would have been culled. There was a burning pyre of hundreds of dead animals three miles away. We could not escape the acrid stench.

It was horrifying for the farm, and just as serious for the ice-cream business. If there was a case of FMD within three kilometres of the farm, then all food and packaging would need disinfected. Effectively we would have had to destroy all the ingredients and packaging stored on the farm, so we took precautions to minimise the risk. We stored all our ingredients and packaging at another food factory in a nearby town. At the start of each week we collected the ingredients and packaging needed for that week, and then every Friday a local frozen food distributor uplifted the ice-cream made that week, so that if we were forced to destroy all food on site, then the loss would be minimised. There was no leeway for late deliveries from the supermarkets. For them it was business as usual, while for us normal had gone up in flames as soon as those pyres were lit.

There were big implications for our team too. A few of the people we employed lived on other farms and, as already mentioned, it was too much of a risk to have someone moving from one farm to another on a daily basis, so some staff had to be laid off. There was no such thing as furlough back then.

A universal message went out from government to say that the countryside was closed. Our visitor centre didn't open until June, and even then some neighbours thought we should have stayed closed for longer. Gift shops, accommodation providers and restaurants were closed. For months we had virtually no income, at what should have been our busiest time of year.

As we pulled ourselves painfully back to our feet post-FMD and looked around at the organic ice-cream market, things had changed dramatically and all the momentum we had built around Galloway Organic hit a brick wall. The growth in organic food continued at a very encouraging rate, we just couldn't respond to it. Instead the big boys spotted our opportunity and they ran with it, sweeping up the organic ice-cream market and pushing us aside.

Over the next year a wave of dairy farmers came through the organic conversion process and suddenly, where there had previously been very little surplus organic milk available for processing, now there was a flood. The big boys could buy this surplus for the same price as conventional milk, put on an organic processing line and fire out organic ice-cream a lot cheaper than we could. In fact, this was the ideal outcome for the supermarkets, because they preferred to deal with big players and known suppliers.

We were left figuring out how to pay back enormous loans on our newly expanded factory with virtually no income from the crucial summer months of that most important year. In the space of two years, we had gone from being one of only two manufacturers of organic ice-cream in the UK, to facing competition from all quarters. It was a tough lesson. If the big boys can do it, they will, and they'll do it cheaper.

Chapter Eleven

David

It was the fourth week in February when the first left-of-field event hit the business. I remember it clearly, because it was the same week that I ran over my beautiful wee sheepdog.

Smudge was nine months old and had been running alongside the tractor as I took a load of logs back from the wood. As I approached the farmhouse my father's old dog, who hated Smudge, rushed out. Smudge dived sideways and went under the tractor front wheel. She ran off howling leaving a trail of blood. I dreaded the worst. Eventually I caught up with the terrified wee thing, cowering and shivering in shock at a farm gateway. I loaded her into the pick-up and took her to the vet.

The wheel had gone over her head and dislocated her jaw. The vet did his best at stitching her up, re-setting the jaw and strapping her up. She survived and was a good natured and intelligent part of the farm team for the next twelve years, though she had an odd depression on her skull from then on.

Then Foot and Mouth Disease (FMD) swept through the countryside with devastating results. A nasty, highly contagious disease of cloven hooved animals, domestic or wild, there hadn't been an outbreak of FMD in the UK since 1964 and the response to it was brutal. Any farm with a case diagnosed by a vet would be locked down, as would all farms within three kilometres. Within a few hours a killing team would arrive on the infected farm and the animals on that farm would be slaughtered. The other locked-down farms inside that three kilometre perimeter would subsequently also have animals slaughtered, but that could be weeks later; weeks of caring for animals that were destined for destruction.

The carcasses of the slaughtered animals were transported to massive open-air funeral pyres and burned. There were smoke columns all around us for months, and the wind would regularly bring the stench of burning flesh to our farm. It was like hell on earth. The rural economy collapsed and many small rural businesses went bust.

As is often the case, the crisis brought out the best and the worst in some folk. Rumours and counter-rumours abounded. There was a story that an individual had bought sheep at the Longtown market sale where FMD had been rife, but had never reported it. In those days the traceability of sheep was non-existent. He had taken grazing lets all over the Wigtown peninsular, not far from us, and those sheep subsequently tested positive for FMD, taking out all the farms within three kilometres of each of those grazing areas. The rumour mill claimed he'd been put in a cell by the local police, not because he had broken the law, but for his own safety because there was a furious vigilante team hunting him down.

By this time our financial wellbeing had become heavily dependent on the tourist industry for both ice-cream sales and for our visitor centre. We were hit really hard, and at a time when we were particularly stretched. Technically, we went bust. Unlike Covid, there were no financial hand-outs. We poured in more of our personal savings, raised some money from the family and re-scheduled our loan repayments.

The farm fared slightly better. At least we could still sell our milk, but we couldn't move any animals. We had to buy in extra food to feed the animals we would normally have been selling in late February and early March. There were many farms in the same position, so a welfare cull was set up to take surplus stock off the farms in lockdown. Since I had been planning on selling our flock of 200 sheep, I booked them in. I wasn't happy but these were exceptional times, I had to do something to ease the pressure and to protect what could be salvaged. The cull payment for the sheep was £100 a head but a realistic market value for my flock at that time was £40 a head. Our ice-cream and tourism businesses were going bankrupt so sending the sheep to the cull would net a valuable £12,000 – money we were desperate for.

In the end the cull was heavily over-subscribed, and they didn't phone us until mid-May to come and take the sheep away, by which time they had all lambed. We didn't have the time or resources to hand raise 250 new-born lambs, we were in lockdown after all, so those lambs would need to be euthanised. I couldn't do it. I cancelled the lorry. We limped through the next few months and in October I sold the entire flock for a much reduced price to a hill farmer in Cumbria.

One of the reasons I had for selling the sheep was that grass production on the farm had crashed by about a third, which meant overwintering and our overall stocking rate were presenting major problems. Many years of putting on hundreds of tonnes of soluble fertilisers and spraying weeds had killed a lot of the clover and had damaged the biology of the soil. When you farm organically, clover and healthy soil biology are vital ingredients for grass growth.

We were beginning to learn – the hard way – that organic farming was a completely different way of managing the land and the animals. Over the years we learned a lot more about soil biology, the needs of our pastures and the pests and diseases of our livestock. Previously we just needed to know what chemicals and drugs to apply and when, like following a set of instructions but without needing to understand the processes. Now we had to understand those processes fully, because it was only by understanding the system that we could change it, to avoid problems and facilitate good outcomes. It was more difficult for sure, but once we got our heads around this approach it brought a lot of benefits. If you can harness the power of nature and have it working with you, rather than against you, your farming becomes more resource efficient, productive, resilient, biodiverse and welfare friendly.

This gradual dawning excited my hunger for discovery. My fundamental belief in the infallibility of conventional science in showing the way forward had not just been undermined, it had been shattered. I didn't know what to believe anymore other than what we, and other travellers on this organic journey, were seeing with our own eyes and experiencing in the land, the soil and the environment around us. This misplaced faith in the absolute integrity of science was encapsulated by a visiting scientist recently who said, 'You have to realise that, on the whole, scientists tend to discover what they are paid to discover.'

It was partly the compelling nature of this challenge that egged me on, but also the experience of other dairy farmers who had given up dairying, leading to a gradual loss of interest in farming itself. Dairy farming is very hands-on. You work with the cows every day, you become attuned to their rhythms and their needs and there develops a real bond between the farm team and a herd of dairy cows, and that's a difficult thing to walk away from. Even though I knew it was unlikely that any of my own children would follow me into farming, I wanted to build a useful, viable, independent farm unit that would be attractive to a farming family. The alternative was allowing it, and the herd, to be subsumed into one of the big industrial dairies like so many before. I couldn't stomach that.

Thankfully FMD never quite came close enough to our farm to

force a cull. It got within four kilometres, one kilometre closer and our beautiful herd of cows would have been shot. Had that happened, there's every chance I would have finished with dairy farming entirely.

Chapter Twelve

Wilma

Immediately after FMD our focus was simply financial survival. The painful irony was that if our herd had been culled we would have been financially compensated – very generously – and not just for the livestock culled, there were jaw droppingly high payments for the clean-up afterwards. So much so that the compensation paid to farmers allowed them to match-fund grants to buy new equipment and build new sheds. For some it was a once in a generation opportunity to modernise their farm.

Meanwhile Cream o' Galloway needed to rebuild ice-cream sales, but we didn't have the money it would take to do that. There were two positives, although they didn't feel like positives at the time. The first was that our bank was understanding. We had a good long-term relationship with our bank manager, whose customers were all suffering from the knock on effects of FMD. Loan repayments were rescheduled, although we were now on their 'at risk' list, which meant regular financial reporting to their head office.

The second positive was that Scottish Enterprise, a public agency tasked with business support, offered us the advice of a consultant. Now even in those days David and I would roll our eyes at the word consultant – of course that's a little hypocritical of me given my early career in IT consultancy. We got John Whitehead, and our first impressions of him were completely wrong. John arrived in a very smart suit, complete with a pocket square handkerchief, while David greeted him in his customary boiler suit. John had a wealth of experience in the food and drink sector in Scotland, but most importantly he had empathy and a good knowledge of farming. He was also the person who challenged us most and made me work the hardest.

He visited us once a month and at the end of each meeting left me with a huge amount of homework to do to determine the viability of every aspect of our business. His conclusion was that we needed income quickly and the only part of our business that could deliver income quickly was tourism. It was also John who, very diplomatically, told me I was the bottleneck in the business. I couldn't do everything myself and if we were to make the most of a projected bounce back in tourism in the local area, then we needed to appoint a manager for our visitor centre. He was right.

In 2002, Helen Fenby moved from Leicestershire to Dumfries & Galloway to take on the role of our Visitor Centre Development Manager. David and I were blown away by Helen at the interview and chose her over local candidates even though we believed she would probably only stay with us for a couple of years. In fact, Helen was with us for fifteen years and she made the biggest single contribution to the development of Cream o' Galloway of any member of staff over the years. She enthusiastically helped to develop our vision for the visitor centre, introduced farm tours and wildlife events, and at the same time she was fantastic with the general public, which is definitely not a comfort zone for David and me. When she left, Cream o' Galloway didn't just lose an employee respected by everyone, I lost my rock and confidante.

With Helen in charge of the visitor centre, I could concentrate on recovering our ice-cream sales, and that was a hard task. The customers in south east England who had been stocking our organic ice-cream before FMD had largely changed to one of the many bigger brands that had recently been launched, so we decided to concentrate on organisations in nature-based tourism like ourselves. Some wanted to stay with local small-scale ice-cream makers, but there were enough who wanted to work with suppliers who had similar environmental ethics to enable us to slowly crawl out of our financial hole.

We still couldn't get any success with the supermarkets. Part of the problem was our brand names – 'Cream o' Galloway' and 'Galloway Organic'. We were learning the hard way that using a regional name in your brand only works if your region is known for that product. So Cornish ice-cream works, and Galloway beef works, but Galloway ice-cream? A complicating factor is that many people in the UK think that Galloway is in Ireland, including a surprisingly large number of Scottish people. We once had the bizarre experience of selling ice-cream on Edinburgh's Royal Mile on a day when there was a Scottish Rugby match at Murrayfield. We lost count of the number of Scottish rugby supporters who asked us why an Irish company was selling ice-cream on the Royal Mile.

We decided we needed a new brand and a new approach. The organic ice-cream market had become very crowded, so we needed to differentiate ourselves. After much research we decided to launch an Organic Fairtrade range. The Fairtrade market was growing fast and it made sense to us that as well as minimising our impact on the environment by using organic ingredients, we should also actively support the people who grow those ingredients.

It took me a while to understand what could be officially 'Fairtrade' and what couldn't. To launch a Fairtrade ice-cream required that any ingredient which can be Fairtrade, must be registered as such. This meant that the sugar, vanilla, cocoa, mango and bananas that we used in the ice-cream must be both organic and Fairtrade. We had worked with a Glasgow-based design company since we launched, so we went back to them to ask them to help us find the right brand name and design. Our Made Fair branding was the result and it was beautiful; little farmers and growers with their watering cans and hand tools tending their crops.

For once we had followed the textbook with this product launch and we were confident we'd done everything right. Market research, extensive product development, focus groups, professional PR, journalists invited to the farm and we recruited an experienced sales and marketing consultancy. However, the long lead time from concept to launch was happening in parallel with a personal awakening about climate change, biodiversity loss and economic collapse. We had started to question what we were doing, in particular the ethics of producing energy-intensive ice-cream.

Chapter Thirteen

David

In September 2007 we holidayed on the Isle of Mull. On the first day we walked and cycled in glorious summer sunshine, the sea sparkled and the scenery was breath-taking. Why, oh why did people go abroad when we have this beauty on our doorstep? The next five days answered that question.

The weather forecast suggested a front would push through and better weather would follow, but it sat above us for the entire week, bringing thick mist and steady rain. There was little else to do other than read the books we had brought with us, 'just in case'. These included the kind of books we had bought because we knew we should read them, but their heavy-duty content made reading them quite an effort. Well, here was the perfect opportunity.

There was one book called *The Long Emergency* by JH Kunstler. In it he wrote about all the impending crises facing humanity, one of which was a section on the American sub-prime housing market which he predicted could bring down the world economy. He'd written this in 2005 and we read it in autumn 2007 just as Northern Rock was crashing into bankruptcy. Northern Rock was the canary in the global financial coal mine. The near collapse of the global banking system and a recession followed.

Kunstler didn't really tell us anything we hadn't heard before – climate change, biodiversity loss, ocean pollution, resource shortages, diet related disease, pandemic, societal fragility, etc. – but he joined the dots in a way that alarmed us. We could easily have dismissed much of it as doomer exaggeration, but he'd spotted the 2007 banking collapse back in 2005. Not many people had. This guy was smart. We realised that should he be right about even half the stuff in this book, we had to change.

What was beginning to dawn on us was the realisation that if we wanted to develop a really sustainable food system we had to use, as much as possible, only the resources we had direct control over. Buying in loads of fertilisers, feed, fuel and pharmaceuticals left us wide open to global fluctuations in commodity prices. With increasing geo-political instability, part-driven by climate change, these items could rocket in cost at any time leaving our model of food production uneconomic.

We realised that we had to move towards a more 'closed loop' approach, using natural processes to replace the agri-tech ones. Technology was fine in its place, but it couldn't be the focus. We would use technology as a tool to help us reach this closed-loop, circular objective, where we are merely facilitators of – and co-operators with – a nature-based food system.

We had a good little business. We were organic, we'd planted thousands of trees, we had insulated all the farm cottages and installed double glazing and solar-thermal panels. We had already gone a long way towards addressing many of the challenges within our sphere of influence. We'd even spent the summer of 2007 constructing an eighty-foot high, 50kW wind turbine on the farm for our local community. That was exciting.

We'd been looking at wind turbines for some time, but they were too expensive – with all our various projects we were constantly cash strapped. However, the local community could access grants so I went along to a community development committee meeting and put forward the suggestion that we could provide the site for a turbine that the community would own, and then we would buy the electricity from them. They would then have seed money for other community development projects and we would have green energy – everyone would be a winner. But there was a problem, they had no money and knew nothing about wind turbines, let alone their construction. To be honest, neither did I.

They batted the turbine ball back to my side of the court. They would be happy to go along with this idea if I did the project management, and applied for the grants, and if I bankrolled the project until the grants came through. It was not exactly what I wanted to hear, but it was an interesting challenge, so I agreed to look into it.

Wilma and I took the car and then cycled the last couple of miles along forest tracks to a recently completed wind farm in the remote Galloway hills. The final stretch was a steep uphill climb. We picnicked under the gently whooshing blades of this giant machine in the afternoon sun and it felt good. It felt right. Of course, ours would be a quarter of the size of this thing but it would be big enough. It took a couple of years to get everything in place.

We worked with an adviser from the Energy Saving Trust who advised us that the best value for money was a Canadian machine. The site for the turbine was 500 yards from the ice-cream dairy. We had to get the connecting cable spec. and a price. What about the turbine foundations? Concrete? Steel? The turbine blades, generator and electrics were coming from Canada, but we had to source the mast ourselves. Eventually we found a gas pipeline manufacturer in Aberdeen who could make it.

Next were the finances. We hooked up two grant funders providing 80% of the capital cost. We had to pay out for the time and materials up front, and only when the thing was built could we claim the grant. So, I had to get a bridging loan from the bank, secured by the promise of the grants.

Then planning permission. Wind turbines were pretty new to the area back then and there was a lot of local resistance to windfarms, but this was a community project and a single small turbine, so it went through planning without a hitch. Gradually we accumulated the various components of the turbine which sat around the farm in their delivery boxes. Finally the mast arrived. It was in two sections which bolted together, and each section weighed about three tonnes. All the parts were here, it just needed someone to put it together.

I phoned the Canadians to check when they were coming over to direct operations. To my deep concern they said the price didn't include installation and anyhow, they were far too busy at that moment. They'd come and commission it once it was built, but we'd have to get on with it ourselves. I replied, 'But we've never done anything like this before. We're just farm boys.' Their response? 'Sounds good!'

The site was solid rock and we had to dig out a four by four by three metre hole. Our neighbour arrived with his 20-tonne digger and spent a fortnight chiselling out the hole. Jim fabricated an intricate framework of reinforcing steel rods which we lifted into the hole along with a huge steel ring with ten, two-inch-wide holes evenly spaced around the ring, into which fitted the ends of giant, eight-foot-long threaded bolts that would anchor the base of the turbine mast. We then filled the hole with ready-mix concrete, leaving about six inches of the giant bolts protruding above the concrete surface. Two large mobile cranes had been booked to come 100 miles from Glasgow in two weeks' time. That would be the moment of reckoning!

We moved all the turbine parts onto site the day before the cranes arrived, which they duly did, and the weather was perfect. First off, we had to lift the mast parts to enable us to bolt the two sections together and then the drive train onto the mast and finally the three, eleven metre long, fibreglass blades bolted to the drive train. That took more than half a day and there were plenty of 'Oh no!' moments, but some quick work with a gas cutting tool and angle grinder sorted the crises.

Finally, the moment of truth. The complete turbine was gradually lifted into the vertical position and swung over the giant bolts protruding from the top of the concrete base. Each of the ten bolts had to fit through the corresponding holes in the base ring of the turbine mast. If it didn't it would take another day to sort the problem and these cranes were costing £2,000 a day! Thankfully the bolts slipped through the holes perfectly.

The rest of the job was fairly straightforward, a hut for the controls, burying the two tonnes of electric cable to connect the turbine to the ice-cream factory and getting that all connected up. We were ready for commissioning. The Canadians arrived from Nova Scotia a few weeks later and we were ready to roll. There was the grand opening, champagne – well, my sister's fizzy wine – the great and the good and it was a glorious day, but there was no wind. Before anyone arrived, we had reversed the polarity to allow us to drive the turbine using power from the ice-cream factory. It wasn't really in the spirit of the thing, but it certainly looked impressive.

Of course, the Canadians weren't going to be able to service the turbine but with a bit of instruction, we figured the farm team could do that job. Jim was supported by Abe, a very likeable ex-soldier and joiner from London who had moved here with his young family 'to get away from it all'. I left the servicing entirely to them. I have no head for heights, though I did climb to the top, twice, just to prove I could do it, which quickly reminded me why I shouldn't.

The turbine worked away for several years but getting spare parts became more and more difficult. The original Canadian company we'd been dealing with went bust. The contract was taken over by a second Canadian company, which also went bust. Ultimately, our patching-up failed to revive the bloody thing and it sat stationary for a couple of years until a turbine engineer from Cumbria spotted it and took pity, offering parts and labour at cost, plus some ice-cream. In the summer of 2019 we had it going again but in November that year we were hit by a ferocious one-off lightning strike that took out the control panel, along with most of the tills and computers in the visitor centre. Despite some parts from the engineer and a bit of maintenance repair work, it hasn't moved since. It now stands there, an emblem of the futility of micro-renewables.

Chapter Fourteen

Wilma

During our holiday in Mull I think it was David who was the first to say 'We're making an unsustainable product.' He was surprised that I didn't argue with him. Once you accept that oil is a finite resource and that we are nowhere near to finding a truly sustainable, consistent energy supply to meet our lifestyle demands, how can you justify, either financially or socially, using fossil fuels to make ice-cream? We may make one of the best ice-creams in the world, but the process still involves buying ingredients like sugar, vanilla and cocoa from all around the world. We then heat those ingredients up with milk and cream, cool the mixture down, freeze it, and then hold it in a freezer at around -25°C in cold stores, refrigerated vehicles, shops and home freezers. Yes, it's great to treat yourself now and again, but that finite energy resource should probably be going to something a little more essential.

We also quickly agreed that the product we should be focusing on for the future should be cheese. The farm had made cheese from the mid-1800s until 1971, when mass, industrial production priced traditional farmhouse cheese out of the market. We realised we could learn a lot about our future from looking at what had worked in the past.

Back in David's father's days, the farm had made 15 tonnes of cheese a year. The milk was taken straight from the cow to the cheese vat without being pasteurised. Once the cheese was made, it was stored in ambient conditions without refrigeration or humidification. There was very little energy used – apart from lots of physical energy – so surely we could do the same again?

At that time we had advisors from Scottish Enterprise who reviewed our business plans, and on the vast majority of occasions their advice was relevant and valued. But when I explained our new ten year plan to change the whole business model, the advisor strongly cautioned against it. We were more or less told that everyone who had been involved with our business over the years thought we were making a major mistake – in fact, they believed we were threatening the viability of the whole business. It was clear that in 2008, climate change, resource depletion, energy costs and environmental degradation were not considered things that a small business should be concerned about.

When we started to put together a detailed plan to change our farm and business models, it wasn't just the business consultants who were concerned about our proposed changes – we were too. Perhaps the best description of our business is that it's a purpose driven one. We wanted to change our farming and business model in order to make societal changes, rather than to make bigger profits. But when you take your eye off the finances for too long, you can threaten the foundation on which everything is built.

Change is not easy. It put a lot of pressure on us as individuals and as a couple. David became focused on the new environmental paradigm while I had to ensure that 'business as usual' continued so we could pay our staff, reduce our loans and plan a future environmental transformation. We would both get frustrated with each other. David would accuse me of spending too much time working on the existing business and I would accuse him of being reckless in not understanding that this would be a long term transition, stressing the need to spend time on those activities that were currently keeping us afloat.

In our less confrontational moments we agreed that we had to launch the new Made Fair range. We had already invested so much time and money in it, and it did, undoubtedly, have the potential to be a successful brand. In fact, we concluded, if Made Fair did become the success we predicted, then I could see there might be potential to sell the brand and use the money to environmentally transition the farm much more quickly.

We launched the range in the autumn of 2007. I spent a lot of that winter travelling to London on the West Coast rail line and we made significant progress. We had one major retailer on the point of listing our range nationally, and we were in negotiation with several restaurant chains, but by that point it was early 2008. Everyone was glued to the TV watching the wheels coming off the global economy. The bankruptcy of banks in the US in late 2007 was now being followed by Northern Rock in the UK. Panic struck all retailers. Almost overnight their buying strategies changed from premiumisation to focusing on own label value brands. Made Fair's major retail listing was dropped.

The range did well in niche shops and restaurants who valued the exclusivity, the quality and the values of the range, but given the effort and the cost behind the product launch, it was a major disappointment for us. I still look at the beautiful packaging and smile wryly at what might have been.

While the UK largely turned their backs on organic food after the economic crash, most of the rest of the world continued to value organic farming. Our only major venture into export markets was to South Korea after we were approached by Mr Lee. It was a fascinating project. Seeing ice-cream parlours in Seoul and Busan decorated with our Made Fair branding, and watching South Korean TV programmes featuring Mr Lee importing this exciting new brand, was a major confidence boost. But to be selling more Made Fair ice-cream in South Korea than in the UK was never the plan, and sending shipping containers of frozen ice-cream to the other side of the world certainly didn't fit with our growing alarm about the climate.

Accepting that the product we had been making for over ten years should no longer be our core product was hard to swallow. Even though we are a tiny company, the process of introducing cheese has been like turning the proverbial tanker, and there's been many a time it has seemed we are having to turn that tanker in very choppy waters.

Chapter Fifteen

David

During the summer of 2008, after our Mull epiphany and the 'lead-balloon' launch of our Made Fair range, we were debating the future of the farm dairy. We simply couldn't continue with it as it was. The parlour and winter housing were small, gloomy and leaky. The silage pits were also leaky, but in a polluting kind of way, and the waste storage was inadequate. We were confronted with the decision of building something new or getting out of dairy farming entirely.

Our dairy herd of, at that time, 75 Ayrshire cows had been moved out of tied byre housing in the mid-1970s. In a tied byre each cow had her own stall and would be tied there with a chain round her neck all winter. Woe betide any cow who tried to pinch another cow's stall in the summer months when the cows were brought in twice a day from pasture – if that happened there would be quite a dust-up!

Dad had put up a large, general purpose shed in the late sixties. In the early seventies I drew up the plans to convert it into a loose-housing dairy, where the cows could wander about freely all winter, helping themselves to silage from a large feed bunker. It was considered better for animal welfare, and I guess it was – slightly. The main attraction was that it saved a lot of time. It did the job for thirty years, but time moves on and environmental regulations were being tightened. Both leaky silage pits and inadequate slurry storage were now a no-no. We could spend a lot of money on it but still only have room for 75 milkers.

We needed to carry at least 100 cows going forward, and even that was considered too small to be viable by the farm consultants. Interestingly, the visitors who came round the farm on the daily, summer farm tours just accepted the conditions as 'normal'. Of course, they had nothing to benchmark it against.

We got an architect to draw up plans, got them costed and a couple of firms to quote. To build a parlour, winter cubicle housing, silage pits and slurry store, with all the bits of machinery needed, would be pushing a million quid. For 100 cows? Ouch!

But there was another question that needed to be addressed. Would we leave the calves with their mothers? I had been coming under increasing pressure from Wilma and our visitor centre manager, Helen, to at least look at some kind of cow-with-calf dairying. With so many people coming to enjoy ice-cream, we had introduced daily farm tours several years previously. Visitors took in what we were doing with little adverse comment, with the exception of one point during the tour. The visit to the calf house.

Chapter Sixteen

Wilma

Back when David and I first met, he showed me round his farm and I have to admit, I was a bit taken aback. Both sets of my grandparents were farmers but nowadays they'd be called smallholders, so my exposure to farms was basically two cows, a few hens and a couple of fields of grass for hay.

David went out of his way to tell me that Rainton was one of the least intensive farms in the area, which made me feel a little better, but not much. In winter the cows were packed into uncomfortably crowded and damp sheds, then spring would come and they would all gallop back out to grass and I would forget about it. When winter came again, I knew what to expect and began to become inured to it. It became normal.

It was the same with the separation of dairy calves from their mothers. My knowledge was from a very low base. Initially I didn't even know there were cows bred for beef and cows bred for milk, as far as I was concerned they were just cows. I didn't know that the cows with calves that you see in fields together were beef cattle, and that the cows you see on their own were dairy cattle, with their calves, separated from their mothers within a day or so, being reared separately in sheds with 'formula' milk.

The farm has always employed someone to manage the dairy herd and milk the cows. David only did the milking during the herdsman's days off. In all my time on the farm, the herdsman's partner had reared the calves, so when David was standing in on the herdsman's days off, I would feed the calves.

Initially I just didn't get it. Why give a calf substitute milk powder when you have the real thing sitting at body temperature in the next door shed? We fed the calves from a bucket with a rubber teat, but it was rare for a new born calf to accept the rubber teat. They would happily suck on your warm fingers, so you had to sneak the rubber teat into its mouth at the same time as it sucked your fingers. I couldn't help but wonder why, over the years, dairy cows had been removed from this vital part of their role, but you soon focus on doing the job as well as you can instead of criticising every bit of the system.

As soon as we introduced daily farm tours to Cream o' Galloway I was no longer the only person questioning cow and calf separation. Just as I was beginning to accept that separating the dairy calves from their mothers was an inevitable part of dairy farming, regular questions and obvious confusion arose from our farm tour visitors, many of whom were mums with young children, many of them with personal experience of breastfeeding.

They asked why were there calves in a shed being fed by a human, just yards away from the shed containing their mothers – the most logical source of calf-food? We learned to 'spin' it. 'The cows are working mums,' we said, 'and we humans are the crèche for the calves'. No matter how often I said it, it was never comfortable.

When asked why we converted to organic farming inevitably there are two different responses. David's response is that three of the women in his life convinced him – his mother who had a keen interest in nature, his sister Susan who raised her family on a self-sufficient croft, and me, someone who had read about the benefits of organic farming, but without any practical experience of it. My answer is quite different.

David needs challenges and he gets bored if he doesn't have something new and interesting to explore. He is constantly studying new theories and looking for new ideas to introduce. My discomfort with cow and calf separation would certainly have had an influence on his decision, after all he knew that my first impression of his farm wasn't positive – and that would have hurt – but no-one could have persuaded David to change his farming methods without him fully investigating it and deciding for himself that it was worth trying.

During the exploration of cow-with-calf farming, my influence in and involvement with the farm actually reduced. Over the decades I had rarely thought about cancer, but in 2008, sixteen years after my lumpectomy, a routine mammogram showed a possible recurrence. They recommended a mastectomy and this time I was in complete agreement. Thinking back to my mother, she had a mastectomy in her 50s and lived until she was 90. I was now going to have a mastectomy in my 50s and the surgeon said I had the same chance of getting it in my other breast as any other women of my age, so from that point on I didn't give cancer much thought.

There was one big decision to be made though – was I going to have breast reconstruction or not? If it had happened when I was in my thirties then it would have been a definite 'yes', but in my fifties? When I was in a settled relationship? Surely I wasn't that vain? Turns out I was.

I have nothing but praise for all the people I have met in the NHS. Well, maybe apart from that first surgeon who so unsympathetically told me that what he thought was a cyst was a malignant tumour, but perhaps you always take a dislike to the first person who tells you that you have cancer.

I had an appointment with the breast reconstruction surgeon to discuss the options on a Saturday at 4pm. When I arrived, it was clear there were still two people ahead of me in the queue. The one immediately before me was a young boy with his mother. The surgeon I was there to see is also a specialist in hand reconstruction, which the young boy needed, and that put my situation into perspective. Now hand reconstruction is vital, I told myself, breast reconstruction is not.

It was 5:30pm by the time my session started, and it was 7pm when it finished. The surgeon was both informative and sympathetic, recommending TRAM flap surgery; effectively a tummy tuck with the fat removed used to form a breast. I am ashamed to admit that the tummy tuck was the decider for me. No, I wouldn't be wasting her time nor NHS resources, she assured me. Yes, she could introduce me to someone who had gone through the same procedure before. I then spoke to that person and she had no regrets. She told me she could wear a swimsuit without embarrassment and even though it took about two years before she could turn over in bed in one movement, she felt it was worth it.

I was lucky that my operation was in November, a quiet time at Cream o' Galloway. I could have been home within a couple of days if I hadn't opted for reconstruction, but in the end I was in hospital for ten days. When I got home, I went on frequent short walks to try to get my strength back, but it took about a year before I could stand up straight. There would be no calf feeding or lifting of sheep feed bags for me for the next year, but I was able to get back to office work fairly quickly.

While David was consumed with plans for the new dairy shed, I spent the next year trying to salvage Made Fair and deal with the impact of the economic crash on Cream o' Galloway. We were concentrating on completely different things and, at times, our different priorities caused tension. There is no doubt that our personalities are complementary. The vast majority of the time we both recognise each other's strengths and feed off each other positively. It's when you start to focus in on each other's weaknesses that the trouble starts.

David is visionary and strategic, while I am a good manager. That makes for a successful team bringing about innovative projects. On the other hand David can be reckless and I can be nit picking. If we let that go on for too long, then we enter a downwards spiral. I can honestly say there have only been three times in our thirty years together where we've let an argument go on for more than a day. One of us usually breaks the ice with a joke, but this was one of those three times.

We both believed that this transition to a more environmentally sustainable business model was essential. It was the way we achieved it, and more importantly the pace at which that change would be made, that was the point of disagreement.

I was providing David with multiple versions of potential cash flows for the farm, and the inevitable decision was that we were both going to have to cash up our pensions to keep things going. At the same time I honestly don't think I really believed that David would do the cow-with-calf experiment once the new shed was finished, because he didn't really believe there was support for it from the general public.

By this time we'd spent years planning the detail of this radical change and when we told our staff, friends and relatives about the plans the feedback was very mixed. The farm staff were 50:50. Some were up for a challenge, and some rolled their eyes. Friends and relatives who were in farming thought we had flipped. Some advised us not to do it because it would fail spectacularly, and others were fearful that if it succeeded then there would be pressure on all dairy farmers to change, and no-one likes being forced to change.

Most of my friends are city dwellers not farmers. Their reaction was more of a quizzical 'but don't the cows keep their calves anyway?' When I explained the normal system there was universal support for our proposals. Throughout this period of developing the new cow-with-calf farming system David was feeling a bit deflated by the feedback he was getting, while I just kept reminding him there was a lot of support from the general public for what we were proposing.

One of the most difficult things for me was to keep our team motivated. We shared our plans with everyone who worked here at Rainton – we showed them the architect's drawings for the new farm dairy, we explained the thinking behind keeping the calves with their mothers, and assured them that adding cheese into our product range would be good because it would reduce the seasonality of the business. It quickly became obvious to them that we didn't have a bottomless purse and that the resources we did have, both money and time, were being consumed by the future business. It was frustrating to me and annoying for them that repairs and maintenance at Cream o' Galloway were no longer automatic. They were beginning to feel second class to an as yet, non-existent business.

In a small business staff aren't just numbers in a profit and loss report, they are friends. You know their family, their hobbies and often what their worries are. Rightly or wrongly, I subconsciously took the decision that the less I talked about the business changes, the better it would be for morale. It was easier to talk about the here and now than to confuse the team with the detail of such an uncertain project.

There were times when I felt split down the middle. During the day all my efforts were on 'business as usual' but as soon as David and I were together it was back to detailed planning towards the new paradigm. It took its toll. In one summer, two key people resigned from Cream o' Galloway within weeks of each other. It was a huge blow for me. There was no way I could spin it; I was genuinely pleased that they had found new challenging roles, but I was shattered by it and everyone knew it. And most believed that the uncertainty of where the business was going, and our lack of focus on the existing business, contributed to them leaving.

Chapter Seventeen

David

The separation of cows and calves on our farm was becoming a problem because our visitors were seeing with fresh eyes what so many people working in dairy farming had stopped noticing. It should have been easy to simply brush aside their concerns, and I did at first, but I have to confess, I didn't like the standard practice of separating cows and calves either.

I found it deeply upsetting listening to and observing any recently separated cow looking back towards the calving boxes, calling gently then cocking her ears, straining to hear a reply. I found I was connecting with her grief at a very deep level, perhaps because, in a way, I'd been there too. I felt it was a topic worth exploring, if only to rule it out.

The first step was to find people who were already doing it. There was a farmer I found on the internet who was weaning calves at three months. I phoned her and found she was selling them straight away through the local market which avoided the calves hearing their mothers bawling. It was saving her calf rearing time and achieving a good price for the calves, but I didn't want to do that.

Then there was Jonny Rider in Wiltshire who had made farming headlines by developing a system where calves from his dairy herd were raised by foster mothers. In his system, older foster mother cows reared the calves from his 400 dairy cows, alongside their own calf, until they were six months old. I phoned and chatted through our idea of cow-with-calf dairying. He was the only other farmer I spoke to who thought my idea might work.

Further searches came up with the Louis Bolk Institute in the Netherlands who were studying the Dutch cow-with-calf systems. I contacted the project leader who invited Wilma and me to visit. The week of our tenth wedding anniversary found us flying to Paris for a few days' break. On our actual anniversary day, we were on a train to Utrecht to visit the Louis Bolk Institute and from there to a couple of farms doing cow-with-calf dairying. It was mid-January 2009, bitterly cold with sleety showers and the bleak countryside seemed to be covered with a foot of water for mile after mile. Perhaps not the ideal anniversary, but Wilma had suffered me for 17 years so this was unlikely to be a deal breaker.

The two farms we visited were small, with 40 or 50 cows, and both were organic. They were weaning the calves at three months, and it was clear that while the system worked for them in terms of saving calf-rearing time, the calves were suffering a lot of stress at weaning, enough to make them ill. But we could see that the idea was workable and with some modification we reckoned we could improve on what we'd seen.

The next challenge was to get the farm team on-board. This involved Jim Haworth, a very fit and able dairy herdsman, and Charles Ellett, our stockman. Jim was the son of a farmer and had been raised the hard way – working on the farm in early mornings and evenings while attending school during the day. He had come through the college system studying agriculture before becoming a herdsman on a 500-cow intensive dairy farm. He and his wife Sarah had come to us because he wanted to be actively involved in supporting his five young children growing up. He was a judo black belt and an advanced rock climber, and he and Sarah home schooled their kids.

Charles, on the other hand, hadn't come from a farming background. He had studied land management at university and had come to work for Wilma six years previously as a countryside ranger at our visitor centre. He had then transferred to the farm after a year to become our stockman. In his early days with us, Charles distance-studied for a master's degree in organic agriculture and his thesis was on, of all things, cow-with-calf dairying. Charles was certainly up for the experiment and Jim was keen to give it a go too.

Just to be sure, we organised another trip to the Netherlands to two cow-with-calf dairy farms, hosted by the Louis Bolk team. Jim and Sarah, Charles, Helen, our visitor centre manager, and my mother and father all went. I think we had been talking about this for so long and everyone was so used to our sometimes eccentric sounding ideas that there was a sense of eager, if anxious, anticipation about the experiment. And, of course, it was just an experiment after all. If it didn't work out, we could quickly go back to what we knew worked. They all came back from that trip saying, 'Yeah, we can do that.' That hurdle was cleared with remarkable ease.

By early 2009 we had the plans for the new dairy, a price tag of about £1 million and a vision for a new way of dairy farming, but how were we going to fund this? We had massive borrowings already, the economy was in freefall and banks were scrutinising all existing loans for risk of default. We daren't go back to them lest they call-in the loans we already had. We certainly wouldn't mention to them the crazy idea of cow-with-calf dairying. We'd have to raise the money from our savings, by cashing our life assurance policies and pensions, by going round the family, from grants and from any third-party loans we could get our hands on.

I guess I saw this as a step-by-step process towards a goal that I knew was highly unlikely to be achievable. It was like a kind of pipedream, a token gesture where we could turn round and blame something outside our control for the expected failure. This was a common theme feeding into my uncertainties, which is why, having proven to my disbelieving self that it can actually work, I now feel real confidence that this system could make a significant contribution to sustainable dairy farming, and sustainable in all the senses of the word.

At that time there was an EU scheme for farm development with up to 40% grant. Problem was, if we applied as the plans stood, we couldn't increase the claim after the project had been approved, so if we were going to do cow-with-calf, we had to budget for that too. So what else might be needed?

We'd been looking at renewables and anaerobic digesters were the 'new kids on that block'. I spoke to a couple of digester firms and they both said something along the lines of, 'If you're going to put up an AD unit, it needs to drive a 150kVa CHP to be viable.' Initially this was a different language to me, but after doing a bit of research it translated into the following explanation. The gas from the anaerobic digester is captured and piped to an engine that has been converted to run on biogas. This engine, in turn, drives an electricity generator which can pump out 150 kilowatts an hour. Not only does the engine produce electricity but it also produces loads of hot water from the water circulating to cool the engine. Hence CHP – combined heat and power.

It sounded great but the numbers were big. Over a million for the AD and then there was either a three-phase connection, or an upgrade to the electricity grid. This could run to several hundred thousand pounds, but what would be the return? All being well, the cost would be paid off in seven years from the savings in electricity purchased, sales of electricity to the grid and the various renewable subsidies. There was another catch. In addition to all the farm waste, we'd need to put in 2000 tonnes of silage. That was almost all the silage we made in a year! Meaning that we'd need to buy in feedstock for the cows, which was crazy! We were losing sight of where we were trying to take the business – to become a closed-loop, self-sufficient, sustainable food system. This was a million miles from that, so it was time for a re-think. The idea of the AD was right, but the scale had to be appropriate to match the waste that was naturally produced by our farm, without relying on buying in loads of extra stuff.

By luck, we stumbled across a small, new-start, AD firm based in Shropshire. They would design something for us which met our needs, and with the promise that it would be easy to maintain. The cost would be £125,000, but we would need to do all the civil engineering ourselves, and source the materials and equipment. They would project manage, supply some of the gear, install and commission. Deal! As it was only a 25kW system, we wouldn't need a grid upgrade. We were getting an up-front grant and therefore we wouldn't qualify for any subsidies on the electricity or hot water produced, but that was ok, it was becoming doable.

I reckon it took us the best part of a year to get the CHP up and running properly. It was a Cummins diesel engine which had been converted to run on the biogas from the digester. Only it didn't. And it didn't help that the company that was to commission the engine went bust before they'd finished the job. The engine was a good enough quality diesel engine, but the conversion to run on biogas was a total botch. Engines that run on gas, apparently, run a lot hotter than diesel. All the pipes, clips, plugs and leads kept melting. Eventually we had to replace everything with parts specified for Formula 1 racing cars, brought over from America.

We got it running fairly trouble free for a year with the help of the local garage owner and a neighbour of Helen's who had spent his career in electronics. It generated 56 megawatts of electricity and loads of hot water. Success! But that success was short-lived. A fault appeared on the electronic control unit screen. Every time the CHP got up to speed and tried to link to the grid it stopped, and the error message appeared.

That was it! I'd had enough. I had wasted too much time on the thing and, to be honest, it was something of a relief. We also had a boiler that would burn the biogas, and it was pretty low-tech. In due course we would install a second boiler at the cheese dairy to utilise surplus gas produced by the AD from January to April. The CHP unit still sits where it started. I think it might even still start, but I don't care. It really is not worth the hassle, a bit like the wind turbine. I call this 'micro-renewable syndrome', where the effort and cost of keeping this technology operating is just not worth it.

That doesn't mean the AD itself was a bad idea, far from it; it's now an essential part of our farming system. By good fortune the Soil Association ran a series of AD events at the farm with professional input from soil scientist Dr Audrey Litterick. Her detailed knowledge of fertilisers was invaluable in pointing us towards best practice in digestate management in order to build soil health and crop yields.

What the digester does is give the bugs an anaerobic environment in which to break down the fibre in the farm's waste to produce methane, which we then collect and burn in our boilers to produce hot water and reduce our greenhouse gas emissions. The product of this process – digestate – is a much more powerful fertiliser than the raw slurry it's made from. Not only that, but instead of trailing back and forth to the fields with heavy slurry machinery causing soil damage and compaction, we now use an umbilical system. This is where the digestate is pumped up to a mile along a flexible 5-inch hose, out to a low-ground-pressure tractor in the field which has a long bar across the back of it with a series of pipes that hang down to the ground, placing the digestate directly onto the pasture.

The anaerobic digester is an integral part of our move to close the nutrient and carbon cycles on the farm. It also reduces our negative air, soil and water environmental impact significantly. In the good ol' bad ol' days we used a splash plate to spray the slurry around in the fields; it got the slurry where it was needed, but it also released a lot of the ammonia and nitrous oxides into the air. As well as being pollutants, they also contained a lot of the fertiliser nitrogen that was, in turn, lost from the slurry.

As if that wasn't bad enough, when the raw slurry soaked into the soil, the soil bacteria started working on it, releasing methane into the atmosphere and using up the soil oxygen. This meant the worms and beetles living in the soil had to come up to the surface to breathe. Waiting for them were hundreds of crows and seagulls. These clever birds knew over decades of experience that the stink of slurry meant a feast! It also meant that we were depleting our soil health and biodiversity. We hardly see a single crow or gull in the field now after the digestate is applied, it's such a contrast.

Getting back to the plans for the new-build farm dairy, an organic dairy farmer friend was sitting in one day as we were discussing the type of parlour we were going to install. We were thinking of a conventional herringbone with ten cows coming in on either side in a group. We were familiar with this in our old parlour but Gavin spoke up, 'If you're thinking to do this cow and calf thing, you'd be better with an auto-tandem.' We'd heard of these things, but no-one had actually seen one and there were none in Scotland. We tracked a couple down to Cumbria and we all went for a look. Yes, this was perfect. Each cow would have her own compartment which meant we could treat each cow as an individual, giving her time to milk, fast or slow, as she preferred, and time to eat her parlour snack. It would also be possible to allow any calves that ventured into the parlour to exit, which wouldn't be possible with the herringbone.

Once we had all the bits of the system specified, we went back out to tender. It was now going to cost one and a half million, an eye watering sum of money and a total that was going to stretch us to the limit, if not beyond. At this time Helen Browning, hearing of our plans, dropped in for a chat. Helen is the Chief Executive of the Soil Association. She's also an organic arable and dairy farmer, with a shop, a pub and online retail business, so she knows her stuff. We explained our financing problem and Helen said she'd put us in touch with someone who might be able to help.

A few weeks later Ruth Layton, a vet who co-founded FAI Farms in Oxford, looked in and as someone whose objective in life was to explore ways of improving the welfare of farm animals, this was right up her street. Their group of companies had recently been floated on the stock exchange, enabling Ruth to follow her passions. She was in the process of setting up a trust fund and asked what we might need to get us over the financial line. Off the top of my head I said, a hundred and fifty grand. She didn't blink.

I should have asked for more, but I was scared of putting her off. You see, when an entire industry – practical farmers, academic researchers and policy people – are all telling you it won't work and that you are crazy for even thinking it might, finding someone of Ruth's status and experience, who believes in what you are trying to achieve, is enormously empowering. Sure, the money would help but the belief that we could actually make this crazy idea work was more important.

Even with Ruth's loan, this would still leave us short – well short – there was still the grant to secure and there was no guarantee of success. Each application was scored and weighted by the Department of Agriculture and compared against other bids – it was to all intents and purposes a competition. The projects that scored highest on a fixed set of criteria got the money and the rest got nothing. In a way, I almost hoped that our application would be rejected. That way I'd be able to tell myself that I had tried. If we were turned down then it wouldn't have been my fault that we couldn't do the cow-with-calf thing; I could walk away conscience clear. This wouldn't be the last time I had these thoughts.

It was the afternoon of Christmas Eve when I answered the house phone. Andrew Johnson from 'the Department' was calling. He had listened patiently to my description of what we were planning with the cows and calves and how I believed, counter-intuitively, it might work. Andrew, on the phone, said, 'David, I've great news, your grant application has been approved!... David? ...Hello, are you still there?'

I think I was in some kind of shock as the realisation that I was going to have to go through with this sunk in. Eventually I replied, 'Yeah, that's great news. Thanks Andrew.' I suspect Andrew, detecting the hesitation, may have guessed what was racing through my mind.

Not all of the items in the application would be grant aided, but overall it worked out at about a third of the cost, but that still left us struggling to raise the remainder. The next break came when our good friends and fellow organic dairy farmers Ross and Lee Paton sold a piece of land and had some spare cash in the bank. I had been explaining our plans to Ross as we travelled to various dairy meetings and was taken aback when he offered a loan to help with the finances. This was high risk stuff, but I didn't hesitate, I took their hundred grand with sincere thanks.

That still left a hole, a quarter of a million hole. We had tried everything except the banks, but drawing the bank's attention to what we were planning would be very dangerous.

I figured this was an experiment and maybe we would qualify for a research grant for part, or all, of the project. We heard about a European research grant scheme called Horizon 2020 which had hundreds of millions of Euros at its disposal. There was a section calling for applications for developing environmentally sustainable food systems. We figured we'd be a shoe-in. We had a few days planned for a break in mid-Wales and, as the deadline for applications was approaching, we spent most of that time putting together our application.

We submitted it with minutes to spare before the deadline and awaited the outcome a month or so later. It was turned down. Again, theirs was a scoring system for various components of the application and if any section failed a minimum score, then the whole plan failed. It turned out we scored highly in every section except the one covering the commercial outcomes. We weren't producing a whizz-bang, techno-widget that generated intellectual property, jobs and growth, so our application was thrown out. Just out of interest, we checked to see who had been successful. One memorable award in this call was for the development of a low emissions cappuccino machine. For our food system transformation to be bounced out of contention by a coffee maker was gutting.

However, we were now on the Horizon 2020 radar and were subsequently approached by four other consortia to join their various applications. Every single one of them failed. We also approached a UK research funder known as Innovate UK but again, what we were trying to achieve didn't fit with their commercially orientated, high tech, technological objectives. Trying to develop a truly sustainable food production system just didn't cut it with the research establishment, it would seem. Their understanding of innovation clearly didn't match ours.

We were discussing plans for the new dairy one day with the farm guys and Jim said, 'I fancy a go at that. I think we could do that.' I was astounded. 'But Jim,' I replied, 'we've never done anything like this! This is on a different scale altogether to anything we've ever done before.' 'Sure,' he replied, 'but it's just a big Meccano set...'

Well, we both knew that was a gross simplification but inside I knew that if Jim could pull this off, then it was the final bit of the jigsaw that would allow us to fund this project and start us along the road to cow-with-calf. Building it ourselves would save us the builder's profit margin of about a quarter million, and so began a mammoth, four-year building project. Every metre of cement, concrete panel and block, steel, timber and GRP; every piece of cow cubicle, mattress and barrier, every nut, bolt, nail and screw, had to be measured, ordered and paid for. It filled our lives, and anyone we could recruit to help, for the next four years.

Chapter Eighteen

Wilma

When David and I announced to the family in 2010 that we planned to start making cheese on the farm again, my father-in-law was delighted. Then he put his hand on my knee and, with a twinkle in his eye, said 'A change of Hell and a new devil!' He knew we were excited, but he also knew tough times lay ahead.

There were lots of reasons why we returned to cheesemaking. During the conversion of the farm to organic production, we had met many great thinkers who were farming in a more environmentally aware way than we were. We had a lot of catching up to do, not just in how to farm organically, but in understanding why stopping the use of agri-chemicals was so important for the future of the planet. The emergencies of climate change, resource depletion, biodiversity loss, pollution and diet-related non-communicable disease were discussion topics that arose again and again. We decided to tackle improving animal welfare through cow-with-calf dairy, while embracing natural systems and introducing low-energy traditional cheesemaking in one fell swoop, addressing all our growing environmental concerns in one radical system transformation. We knew it wouldn't be easy.

The first step was to learn how to make cheese, and that was my task. The go-to place was Kathy Biss near Plockton, in north west

Scotland, and I went on a three day cheesemaking course in March 2012. Kathy has been making cheese all her life and is called upon for advice by many cheesemakers, from as far afield as America. The drive north gave me time to think about it all. Starting to make cheese on the farm again was a major step, not just on a practical level, but also on an emotional level. Twenty years earlier when we had started to explore how we might diversify the farm, David had been adamant that we wouldn't be making cheese – it was far too much hard work, he said. That, coming from a farmer, seemed odd, because to me it simply felt like we were closing the loop.

On the training course we had to choose two different types of cheese to make. I chose a soft and a hard cheese. One of the other people on the course was Richard Comfort, a restaurateur from the Black Isle, who wanted to start making cheese himself. I was intrigued when he chose a blue cheese. He told me his reason – there was, in his opinion, a shortage of good Scottish blue cheeses. That caught my attention, and I resolved to train my taste buds to like blue cheese if that was the case.

Kathy talked us through the equipment we would need, which depended on the volume of cheese we would be making. Her own dairy had a variety of very small vats – right down to fifteen litre buckets. If all we needed were a few buckets to learn the trade, then in theory we could start as soon as I got home, but we were fast approaching Easter. The tourism season was starting and my focus had to be on ice-cream in the meantime, so it was October before we started experimenting with cheese at the farm.

Richard and I stayed in touch. He had accumulated some cheesemaking equipment, but didn't need it all, and he offered to loan me some. So that October, I arranged for Kathy to come to the farm from Plockton, and for Richard to bring his equipment from the Black Isle, and Kathy gave us a two-day onsite cheesemaking refresher course. Sarah, the wife of Jim our herdsman, had shown an interest in becoming a cheesemaker, so she attended too.

It was during these two days that I accepted I would never be a particularly good cheesemaker. I watched as Sarah and Richard were completely in tune with everything Kathy showed us, while I struggled with taste, smell and feel. Give me tools and a task – a thermometer, pH meter and a clock – and I'm at home, but ask me to wait until the curds split cleanly or look like a chicken breast, then I'm lost. I've now accepted that while I may be logical, I'm just not a crafts person.

The only cost effective way for us to start making cheese was to use the existing ice-cream dairy, which already had much of the equipment we would need. We spent quite a bit of time with our Environmental Health Officer (EHO) showing her our risk assessment for the introduction of cheese into our dairy. The one thing she insisted on was that we would have to pasteurise the milk used to make cheese. We would need many months of records of good microbiology results before she would even entertain the idea of us making unpasteurised cheese. We didn't argue. We just wanted to get our cheesemaking off the ground.

We have always had a good relationship with our EHO, but it was still a surprise when Audrey phoned us 18 months after we started making cheese to ask if she could bring another EHO with her to visit. The other EHO only had experience of large scale cheese factories and a new small scale operation was being set up in his area, so Audrey wanted to show him what a small scale pasteurised cheesemaking operation looked like. We were still learning our trade and smirked at the very idea of us being the cheese dairy they chose to visit as a 'good example'.

Inevitably what can go wrong, does go wrong when those in authority are around. We had agreed with Audrey that they should arrive at 10am, there was no point in them arriving at 7am to see the milk being collected from the milking parlour and then being pasteurised – that's just a lot of standing around – they could just arrive when the real cheesemaking work got going.

The morning of the visit, the boiler that heated the water to pasteurise the milk just wouldn't work. We tried everything, including getting David and Jim to try to fix it. They were flummoxed, we'd need to call in the heating engineer, but we had to get the milk pasteurised before the EHOs arrived at 10am. What could we do with the milk that was already sitting in the pasteuriser? We couldn't return it to the farm's bulk tank. We didn't want to put it down the drain. There was only one solution. We were about to make our first batch of unpasteurised cheese under the supervision of two EHOs without them knowing!

When they arrived we were cleaning out the pasteuriser. We walked them through the whole process and, fortunately, they left without asking if we had actually pasteurised the milk that was in the cheese vat, which meant that we didn't need to 'fess up. My main memory of that day was of Sarah whispering to me every time I was near her. The texture was different; you could actually smell the cows and the curd had much more flavour. She was struggling to hide her excitement.

This batch of cheese was never sold. There were plenty of us on the farm who wanted a piece of this one-off batch, and our decision was unanimous – we were going to make raw milk cheese. We made our first official batch of raw milk cheese a full year after our accidental experiment. The reason for the year delay? That's how long it took us to prepare the volume of evidence needed to demonstrate that we truly knew what we were doing, and that we could be 'trusted' to make unpasteurised cheese.

Chapter Nineteen

David

In August 2012, after four years of planning and building, we were ready to move the cows into the new parlour. We had organised a small opening ceremony and invited the farm staff, a few friends and suppliers of materials for the dairy. I did a short presentation and we had some bubbly.

One of our friends mentioned to Wilma, who had been away that day, that I seemed a bit distracted. I have to confess I barely remember any of the details of that day and found myself wandering about the farm later that afternoon struggling to differentiate fact from fiction and reality from hallucination, and it wasn't the bubbly! My stress levels were through the roof and in truth I was probably suffering some kind of breakdown. Fortunately, the next day I was much better if a bit fuzzy. I guess we never really know how close we are to the edge of sanity until we step over the boundary.

At first the cows were a bit afraid of the new parlour, but within a few days most of them were coming in happily, so we carried on and began the first stage of our cow-with-calf experiment. I've mentioned some of the drivers behind this – Wilma, comments from visitors and my own experience of being forcibly separated from my children – but there was another factor at play.

Over the preceding ten years of farming in an ecological, nature-based way, we had seen crop yields crash but then recover. Our livestock numbers and our productivity returned to pre-organic days, but without all the toxic chemicals we had been pouring onto and into our soils, crops and animals. Regaining productivity while having stripped out a lot of the inputs – and consequently a lot of the costs – had led to our most profitable years in farming. As someone who had previously believed deeply that modern farming was based on good science and data and that organic, agroecological systems were a 'muck and magic' con, my belief system had been turned on its head. Completely.

This led me to question every aspect of our farming systems, including the policy motivations for enticing farmers down a technology-based route. I had found that facilitating and enabling natural systems had, eventually, delivered good results and excellent productivity, so I was deeply curious about whether we could do the same with livestock productivity, and I was excited by the prospect of what we might discover. By enabling natural behaviours, could a cow-with-calf dairy system lead to health benefits and productivity gains too? We were about to find out.

We figured that since the calves would be drinking their mums' milk they'd be thriving and healthy. The cows would be delighted that they could keep their calves and would be calm and relaxed, and of course, they would show their appreciation by sharing their milk with us. Hah! We had a lot to learn, and of course we had to learn it the hard way.

To our huge disappointment the cows did not show their appreciation for the new arrangement. Far from it. In the past, when the cows had come out of the milking parlour on the first milking after calving, they found that their calf had been taken away. But not now. They didn't know the rules anymore. They knew that separation was the norm. Would her calf be taken away today? Tomorrow? The cows got stressed and withheld their milk. Some got mastitis and some went lame. There was less than half the milk there should have been in the milk tank. We went from once-a-day milking to twice-a-day and got a little more milk but really not enough to justify all the extra work. The calves began stealing milk from other cows who would, naturally, object, resulting in cuts on some of the cows' teats from the razor sharp calves' teeth.

There was more bad news. Normally we expect to get milk with a butterfat content of about 4.5–5%, which is perfect for making cheese. The cows who were suckling their calves were giving us milk of 1.8% butterfat which is useless for cheesemaking. This low butterfat milk is the foremilk. It has normal protein levels and is the milk the cow cannot hang onto. To get the high butterfat milk requires the cow to release certain hormones, like oxytocin, which then relax the tiny compartments in the udder holding that high fat milk. This process is called let-down. If the suckling cow is stressed, she won't release those critical hormones to release that milk fat. Apparently, it's the same in humans.

We were stunned and incredibly disappointed. This couldn't go on. We needed milk and there was almost nothing in the milk tank. We decided we would have to separate the cows and calves at night so that the cows would have to give us some milk in the morning. They couldn't hang on to it all, surely?

They did not like that either. We tried feeding the cows at night at the feed barrier where, when set in the 'lock' position, if they put their heads through to eat the feed, they got locked in. Some of the cows ate and were locked out of the way. Others wouldn't put their heads through and fussed around the calves, attempting to stop us moving the calves into the calf-only area in the middle of the shed. Even though it was right beside the cows, in full view and only separating them by a couple of rails, the cows protested very strongly. To make matters worse, the calves quickly sussed the system and would race around the shed, dodging our efforts to corral them.

All that was bad enough, but there was worse to come. We have the herd split into two roughly equal calving groups – autumn (mid-October to mid-December) and spring (April to May), and it's important that the calving is kept tightly grouped for ease of management of the cows and calves. This requires that the cows return to the 'bull', known as oestrous, or more commonly coming into season, in good time to allow them to calve again in the same group the following year. At that time, we were only using artificial insemination so that we would be able to produce three-way crossbred mongrel cows that were naturally healthier, more resilient and productive, but without the complications of having to manage several different dairy bulls on the farm.

It turned out the cows were exhibiting a feature which is also seen in humans called lactational anoestrous. Suckling was a natural contraceptive, causing a delay in the cows returning to be ready to get back into calf by about 35 days. Luckily there was a natural solution to this – have a bull present – but that wasn't much use for this group of cows. More than half the cows in the autumn group failed to get back into calf and slipped through to the spring calving group which meant that not only were we losing almost all our milk to the calves now, we were going to lose milk production the following winter too. We have a 'closed herd' for disease control purposes, which meant we couldn't just go out and buy more cows.

It began to dawn on us that we had done a lot of things wrong. Cows like a routine, and once they know a routine and understand what the rules are, they settle down and are happy. What we were doing throughout this pilot was constantly changing the rules, and the biggest change, of course, was ending the separation of the cow from her calf.

Always before, 24 hours after calving, the cows would be ushered into the parlour, often unwillingly. On leaving that first milking they would join the herd and though they might call out, they knew the calf would be gone. Now their calf was still there. They couldn't understand what was going on. When would their calf be taken away? They had always associated the milking parlour with the removal of their calf. They expected it would happen, but they didn't know when, so they became very stressed.

If the cows were stressed, we were worse. It was all very depressing. No matter what we did we couldn't seem to crack the system, we were getting nowhere and we were rapidly careering towards bankruptcy.

The ewes were about to start lambing and, as the shepherd, I couldn't manage the lambing and help with the cows and calves too. I lost a lot of sleep and suffered delusional episodes again. It was mid-March when, exhausted, I said to Jim, our herdsman, 'Take the calves off the cows Jim, this isn't working.'

We walked away with a strange sense of failure, tinged with relief. The dairy industry was right; cow-with-calf dairy would never work. We had tried our best, my conscience was clear and a huge cloud lifted off my mind.

Chapter Twenty

Wilma

The harder we tried to find solutions, the clearer it became that neither of us wanted to give up. It was a mix of wanting to make the system a reality, and simply not wanting to be seen to have failed. We wanted this painful experience to be a step towards getting a system in place that would work long term. So we didn't give up the trial, we 'suspended' it.

Finances aside, there was a lot about the cow-with-calf pilot that had worked. The calves were growing well and many of the things we had feared just didn't happen. Dairy cows are pretty docile, certainly when compared to beef cows, but the widespread belief is that beef cows are more aggressive towards humans because they have their calf with them. We were frequently warned that our dairy cows would become more aggressive towards us if they were with their calf, but it just didn't happen. In fact, their temperament was much better than we had dared hope.

I can remember at Christmas when David's daughter Margaret, a vet, had bought the calves gifts. She called them environmental enrichment tools, but most folk would know them as space hoppers, and we bounced them hopefully into the calf creep area for the calves to play with. Then out of the corner of my eye I saw Meg, the farm pup approaching. I braced myself. This could go very horribly wrong very quickly, but no. Thankfully the calves saw

Meg as another new plaything and ran up and down the shed with her while the cows just kept a watchful eye. There has never been a safety issue about us or the farm dogs being in a shed or a field with the cows and calves.

However, as was pointed out to us by Ruth Layton, a specialist in animal welfare, gimmicks such as toys are a poor substitute for good welfare because cows and calves together in a herd will naturally create the community that they need. We very quickly found that the calves got bored with whatever new 'toy' we gave them in our attempts to distract them from suckling, and the space hoppers didn't even last a week.

Financial modelling for the business has always been my job. With my IT background I confess that I like to relax with a good spreadsheet. We started the initial pilot to keep the cows and calves together in 2012, and before that we went through all sorts of financial planning scenarios. Let's be honest, at this stage it was all guesstimate.

It's funny how some things stick in your mind. During the early planning of the new dairy system we were about to go to Inverness to visit friends one weekend. That afternoon I had just finished tweaking a spreadsheet with our 'most likely' estimate of milk yields. Amazingly I'd projected that in year three we would be making a profit and managing to pay off the hefty loans that would need to be secured before the project could begin. David and I spent the first half of that five hour journey trying to find the mistake in the spreadsheet. Deep down we genuinely didn't believe it would be financially viable so quickly, because when you know that the income from your milk sales is going to be cut in half, how on earth can you make that back?

Beef has always been part of our dairy farming system, and we have always seen value in rearing the bull calves on our farm, both from a compassionate point of view, and an economic point of view. For much of the dairy industry bull calves are a 'waste' output of a system that revolves entirely around milk production, but nature doesn't work like that. Even 'sexed' semen – selecting for female calves through artificial insemination – is never guaranteed to produce a female calf.

For many years we had raised the bull calves from our dairy system to maturity on our farm, reserving a few for beef sold as burgers in our visitor centre, with the remainder sold through conventional routes to market. The alternative to raising dairy bull calves for beef is shooting them at birth, or exporting them overseas to countries where there's strong demand for veal. David and I have always agreed that those practices are abhorrent. For dairy farming to exist, the bull calves must have a purpose. Our new farming system, where the calves are suckled by their mother, would move us much closer towards a calf from a traditional suckler beef farm, producing superior quality beef to that of typical dairy beef systems. Using an Aberdeen Angus bull, a beef breed, as sire over most of the herd would, in effect, produce a beef calf from a dairy system. For this new model to work financially, the bull calves would become more important than ever before.

The crucial factors were the price we would get for the milk and beef and, of course, the quantities. So few people were operating this type of system that we could only guess at the numbers. There was very little research on cow-with-calf dairy, and there wasn't much on once-a-day milking either. We spoke with a couple of smallholder farmers who were working with between five and twenty cows to find out what they were actually achieving. The data varied markedly and, not surprisingly, so did the predicted profitability of the farm when we fed all the alternative yields into the computer model.

Prices can, of course, be highly volatile for milk and beef, and you can be sure that prices for both are rarely high at the same time.

We needed to secure a stable, predictable price for our milk and that was why it was so important to start making cheese. The aim was for all the milk to be sold from the farm to the cheese business at an agreed, predictable, fair and stable price.

We've been long enough in this game to know that most business plans are too optimistic – sometimes even the ones you think are pessimistic – but even our most pessimistic models didn't anticipate the financial haemorrhaging that took place during that first trial.

Chapter Twenty One

David

With the cow-with-calf pilot suspended, a weight had lifted from my shoulders. Of course, the calves and their mums had bawled at the separation for a few days, but they resigned themselves to the new reality. The March and April calving cows didn't know any better, so life in the dairy returned to how it was before. The new dairy shed was a vast improvement on the old one, so no regrets there. The only reminder of 'the experiment' was the existence of the calf-creep areas in the new dairy. What to do with them? Turn them into cow cubicles and increase the herd to 300 as advised by the dairy consultants?

March 2013 had been glorious weather but the 500 ewes we lambed outdoors had not been getting on with it. Only 60 had lambed by the 22nd. The reason? They had been too comfortable. The odd wet day would normally trigger a splash of lambs, but there hadn't been any adverse weather at all and we had loads of grass. I remember thinking it had been a particularly benign winter and no matter what it might throw at us now, it had been a short one too.

I remember noticing a weather report showing that a cold front would be passing over with snow on the leading edge. Might be a horrid spell, I thought, but snow rarely lay for more than a day or two these days, especially this close to the coast. Sure enough the snow arrived at lunchtime, but then the wind strengthened, with bitter-cold arctic winds driving the blizzarding, powdery snow into huge drifts. This was turning nasty.

We combed the field for new-born lambs in a fog of driven powder, understanding that the severe conditions could render the lambs frozen stiff within minutes. We have numerous little sheds scattered around the lambing field and we'd lift the lambs and coax their mothers into these lifesaving shelters. As darkness fell it forced us to give up and seek sustenance and sleep, we'd continue at first light in the morning.

Against my advice, Wilma had gone to a funeral that afternoon and on her return she had abandoned the four-wheel drive about three miles from home, leaving it stuck in a deep snowdrift. In her funeral clothes, she had battled her way back to the farm on foot. Though she had her phone with her, she didn't call me for help, partly because she knew I'd be up to my eyes at the lambing field, but probably mostly because I'd been proven right. Her steely determination could have cost her dear that night. A couple of days later she went back to look for the car, a neighbouring farmer was clearing the road with a digger but was unsure where our car was. Eventually they spotted the top of the roof-mounted car radio aerial sticking out of the snow. The drifts had completely buried the car, and then some.

We returned to the lambing field the next morning on foot, as snow drifts had blocked the road rendering the farm quad bikes completely useless. Fortunately the biting east wind had eased and the sun was out. The lambing field is a rugged, gorse covered hill falling away on all sides, which offers shelter from most weathers. The blizzard had come from the east – an unusual direction – and on the steep western slope there was dense gorse. Seeking shelter on those western slopes, beneath the gorse, were the ewes. The drifting snow had piled on top of the gorse canopy, pushing it down on top of the ewes and trapping them. We were out of the cow-calf frying pan and into the lambing-blizzard fire.

It took us most of the morning to untangle the ewes from the snow-gorse trap. Back at the farm it was all hands-on deck to clear out sheds and set things up for moving the whole flock indoors. It was the only time we have ever needed to lamb indoors. The ewes wouldn't walk through snow of more than a few inches deep. They were strangely quiet, understandably traumatised and in shock. We cleared the snow drifts off the road with a digger and by evening we were ready to accommodate them in various vacated farm buildings. I walked in front of them with a feed bag and they meekly followed. Triggered by the stress, that weekend 150 of them lambed. It was absolute chaos.

Whenever we had a chance, we returned to the lambing field to rescue ewe and lamb survivors. Tragically the ewes and lambs had sought shelter in the lee of walls and gorse where the powdery snow had accumulated. We took long poles to prod the drifts in search of live animals. We found a few but lost ten ewes and about 60 lambs. It was absolutely heart breaking.

Arctic winds continued through April, and even by the second week of May, when we were sowing grass seed, we were having to work around lingering snow drifts. At least – I told myself – we didn't have to contend with the bloody cow-calf pantomime at the same time. What an awful winter it had been.

Chapter Twenty Two

Wilma

I've always seen myself and David as pretty steady characters, never getting too excited, nor too down, about anything, but the cow-with-calf pilot drove us to the absolute limit. Various comments have been made to me about the Finlays over the years. In the early days, when we were first diversifying the farm, an older farmer said to me 'The Finlays were aye different'. I took that as a compliment, whether it was intended as one or not. But one recurring theme, from friends and relatives who have known the family all their lives, and definitely mean well, is that the Finlays have more brains than is good for them.

As someone who values intellect so much this was a strange notion for me, but there was plenty of evidence of mental health issues in the family. David has five sisters, two of whom have had problems in their lives. The six siblings have had 19 children between them, now all aged between 21 and mid-40s, and we are currently averaging one person per family who has experienced notable mental health issues. In addition, there are several who are simply opting out of today's 'normal' society, but I genuinely never expected David to have such issues.

That changed the day he held an open farm event, inviting interested farmers to see everything we were doing. I wasn't even on the farm that day, but Margaret and Christine, his two daughters, came to support him. When I returned from my meeting, David's event was as good as over. Margaret and Christine said it had gone well, but when I caught up with David a couple of hours later he was delusional. He said he didn't know what was real anymore. He couldn't remember what he had said throughout the day. He was in tears.

I had almost no direct experience of dealing with mental health issues, but from the experience of others within the family, I knew that an extreme episode can change you for life. All I could think to do was run him a bath and sit on the edge talking to him. I phoned a neighbour whose brother had attended the open day. He commented that he was worried about David, and said he had been obsessively rambling. It took weeks for David to get back on a relatively even keel. David still has a tendency to obsess about things and gets very frustrated when others can't see the world the way he does. I worry about him.

Despite the failure of the cow-with-calf pilot we were committed to traditional cheesemaking, and I was determined to do whatever I could to make the best possible product from our organic milk. In October 2015 I received an email out of the blue from a Spanish man who wanted to come to the UK for a few weeks so he could improve his English. He said that if we gave him board and lodgings, he would help us on the farm as needed. His email concluded by saying that he designed cheese dairies. It was like a gift from fate and it took me all of ten seconds to reply!

Andres was fantastic, in fact his four weeks with us was utterly transformational and we hung on his every word about the best conditions for making and storing cheese. We had been on the point of creating a cheap and cheerful cheese store using six old shipping containers. Andres was horrified. He said that if we insisted on going down this 'easy' route, then he would go back to Spain and get some winter sun.

Instead he went round all the redundant farm buildings and he fell in love with an old barn, built in 1787 as a water-powered threshing barn. It's built into a rock, with an upper ground and a lower ground level. He advised that the cheese should be made on the upper level and then stored in the lower level. Cheese used to be aged in a cave or a cellar, where the temperature varies little throughout the year, usually maintaining around ten to twelve degrees Celsius – ideal for cheese storage. Caves are also naturally very humid – again ideal.

So the building we walked past every day and which we had always thought of as a dank black hole had merely been waiting for Andres to arrive to bring it back to life. It would definitely be more expensive than shipping containers, but once converted the running costs should be minimal. No electricity would be needed to keep the temperature and humidity at the right level – the natural thick stone would do that all by itself – so this approach also fitted with the climate friendly transformation we were seeking to create. Andres then sat at our kitchen table designing the cheese dairy layout on his laptop.

When he was leaving we offered Andres money for all his work. He refused, we insisted and he relented on one condition – that we agree to go to Andalucía the following year so he could show us round some small scale cheese dairies he had helped design. Andres knew how rarely David left the farm and how we both try to avoid flying. He was teasing us, but the following May we were on our way.

Andres is a very special person, completely dedicated to small scale, ethical food production in all sectors, and particularly cheese. We already had a great deal of respect for him from the time that he stayed with us, but after seeing him working in his home territory we were in complete awe of him.

In five days he took us to eight cheese dairies, an olive oil maker, a miller and a wine maker – all small scale. We met his friends and extended family and experienced some of the best food we have ever eaten, much of which was grown and made by those around the table. At that time, Andres and his college friend Viktor planned to set up a cheese dairy over a three week period. This was ambitious at best! Our visit was during week four, and inevitably there were still a few tweaks required. On top of being our perfect host from early morning to late evening, Andres was working on the new dairy until at least 1am every night! We were inspired but realistic in our own planning. At that time we thought our cheese dairy would take a year to complete. It actually took more than three.

Chapter Twenty Three

David

With the cow-with-calf pilot now well and truly behind us, in early 2014 my daughter Margaret unexpectedly threw us a curveball. She had been studying to be a vet at the Glasgow Vet School and one of her professors – David Logue – said he was interested in studying the effect of welfare on the health and productivity of dairy cows. She told him about our expensive experiment and mentioned that we had gathered a lot of data, which we had abandoned in a box of files to gather dust. He wanted a look at that data, so he sent down a post-grad student to review it, collate it and write it all up in a report.

The research student wasn't at all interested in our project, in fact, his ambition was to be an in-house vet for one of the big American mega-dairies, so we really weren't expecting much from his analysis. His final report therefore came as a shock. He concluded that cow-with-calf dairying might work. Not only that, but he got quite caught up in the idea and sent through some useful thoughts and information on how to separate cows and calves overnight outdoors, quickly.

Early on we had realised that for this method of dairying to work for us, we needed to separate the cows and calves overnight from about three months of age. There were several reasons for this. Up until three months, the calves couldn't drink all their mothers' milk which meant that even if the cow was a reluctant sharer, there would be some milk for us. A cow's milk production peaks around day 60 after calving and thereafter begins a slow decline. The calf's ability to drink milk rapidly increases and a calf can drink 25 litres a day from month four. The separation, albeit only by a fence or a couple of rails, begins the process of breaking the emotional cow-calf bond and strengthens the group peer bond. Finally, when a calf can get *ad lib* warm milk from their mum, they don't eat much solid food, and this stops the calf's rumen from developing. For the calf to thrive post-weaning they must be able to digest pasture, which only happens in a well-developed rumen.

Cracking this overnight separation in a low stress way was an important step in making a cow-with-calf system work, but to be honest I wasn't keen to go back to that experiment again. The herdsman, Jim, was even less keen. 'It ain't broken so why are we trying to fix it?' he complained. I sympathised, but...

When I had been walking through the dairy shed at night during our pilot, experiencing the calm and the serene contentment of the cows and calves, I'd had a deep sense that, if there was any possibility of making this work – any possibility at all – then we had to try. The window of opportunity was closing and I knew that if we didn't do it now, then we'd never do it. I'd be 62 at my next birthday. No regret is greater than the regret of not trying. This was a last chance to solve the cow-with-calf challenge.

Undoubtedly the dairy industry had been purring with self-satisfaction that this troublesome disrupter had been put back in their box, but I had become so disconnected from the dairy mainstream that I barely noticed. Certainly in our, by now very infrequent, conversations with other dairy farmers cow-with-calf was a glaring omission. I also suspect our financiers were somewhat relieved that there was now a good chance they'd see their money returned, with interest. Well, that was about to change.

In October 2016 we started the cow-with-calf dairying experiment for the final time. Wilma and I decided to embark on a three year trial across the whole herd. We would either figure out how to make cow-with-calf dairying work, or we'd quit.

Chapter Twenty Four

Wilma

While I had never really expected the original cow-with-calf pilot to go ahead, after we had pulled the plug on it I was certain that David would give it another try. He still maintains that in those intervening years he had no intention of going back to it, but he never gave me that impression.

The capital investment had been huge. We'd gone to everyone we could with cap in hand and we'd both cashed in our pensions to build the infrastructure that was needed. The pain of the failure, both emotional and financial, was immense, but we could only show that pain to each other. We had to maintain the positive 'lessons learned' message to friends, family and staff. There were only a handful of non-farming friends that we could share our extreme disappointment with.

In the following years, after we suspended the trial, the farm was the most profitable it had ever been. David had to make a decision. Now that the big black financial hole was starting to be filled in, would he go back to the challenge that had consumed him for years, or would he opt for security and stability?

We have completely different recollections of these years. Perhaps we have both built narratives that fit our personalities. I've lost count of the number of times David has said that he had to be dragged kicking and screaming back into the project. I try really hard not to contradict him when he's in full flow and citing the people who – he says – persuaded him to carry on, myself being one of those people. That's not how I remember it.

Unquestionably, we were both very nervous about re-starting the experiment, but my recollection is that everything David said to me privately was about what he would be changing to make sure the experiment worked the next time. Even in our darkest moments we were still trying to figure out how to make it work, rather than debating whether we should walk away.

It was only when we were on the point of restarting the experiment in 2016 that I managed to persuade David to visit those who had experience of doing cow-with-calf successfully. The go-to person was Fiona Provan of the Calf at Foot Dairy in Suffolk. Fiona had started around the same time as our original trial in 2012, but she had stuck with it. Fiona had a herd of about twenty Jersey cows, who she called in to be milked one at a time by their names. Her farm was jaw droppingly impressive. The gentleness and kindness within Fiona's approach was very obvious. The calf could choose whether to come in, but even if it didn't, the cow could still see her calf and Fiona had no problems with her cows withholding their milk.

Another farm we visited was the wonderfully named Smiling Tree Farm in the Welsh borders. Christine Page has a beautiful farm and runs a pristine operation. Like Fiona, she has been extremely generous with her time when others approach her for advice on how to start up a cow-with-calf dairy. One thing you can say about all the people we have met who are running dairies that keep the calves with their mothers is that we are an incredibly determined lot, some might even call us stubborn. We do what we are doing because we believe it is right.

There are lots of small cow-with-calf dairies in England, but only three that I know of in Scotland. One of the main reasons is that it's illegal to sell unpasteurised milk in Scotland, but not in England and Wales. So Fiona and Christine can have a small herd of cows and sell direct to the public, but they don't need to buy a milk pasteuriser in order to do that, they can simply bottle the milk. Meanwhile in Scotland, milk being sold to the public needs to be pasteurised, then cooled quickly and then bottled. To justify the cost of the pasteuriser, you need a lot more cows, and when you have a lot more cows you have a lot more milk, so you need bottling equipment too. The capital cost to start-up in Scotland is huge.

While I never had any doubt that David would go back to cow-with-calf dairy and make it work, there were times that even I thought we should abandon the whole thing. There can be no worse feeling than realising that, despite your very best intentions, at times you have been doing more harm than good.

Chapter Twenty Five

David

Our trial in 2012 had been with 37 cows and calves. Now, in the autumn of 2016, it would be the entire herd of 100 cows, who were split into roughly equal autumn and spring calving groups. We had a fairly clear plan of how we would manage the routines. This time there would be no chopping and changing, which should help the cows quickly settle into this new system. The routine would reduce stress, which should stop the cows from withholding their milk. The plan was to leave the calves with their mothers 24/7 for the first two to three months, and to separate them overnight thereafter. The overnight separation would help the calf's rumen – the stomach – develop, allowing them to properly digest pasture, and it would make sure we had milk in the tank to turn into cheese.

For the autumn calves, we altered the internal layout of the dairy shed to allow more internal divisions and feed areas, which gave us greater flexibility. It also created a special calf feeding passage and a barrier in the calf-only area. The calves would have their own grooming brushes in their calf areas, and we would have gates strategically placed to stop the calves running circles round us once we came to overnight separation.

For the spring calves, we installed fenced cattle tracks with shedding facilities for overnight separation when the cows and calves were out at pasture. These gateways off to the side of the track had a low bar at the entrance into the calves' paddock that allowed the calves to enter but was too low for the cows to get under. That would make overnight separation easier and, therefore, much less stressful, for both the cows and calves, and for us.

This overnight separation was the most obvious change from the 2012 pilot, but it wasn't the only one. What we had come to realise was the importance of looking at the herd as a biological system. Cow-with-calf dairy farming isn't just about leaving cows with their calves. Sure, that's the icing on the cake, but to make it work required a complete system change. One big advantage this time round would be our mindsets. We were more prepared. We knew what was coming and we knew how we'd deal with it.

We'd brought on one of our home-bred Aberdeen Angus bull calves and the vet had vasectomised him. To all intents and purposes, he was a functional bull – all the hormones, pheromones, actions, the works – he just couldn't sire any calves. The vasectomised bull would sort the delayed oestrous, because his presence in the herd should correct the contraceptive effect of suckling.

A major draw-back in our 2012 trial had been the once-a-day milking of the first time-calving heifers. Once-a-day just wasn't enough to get them used to parlour routines, so we started giving them parlour training for a couple of months before they calved. Basically, this meant bringing the heifers into the milking parlour and letting them get used to the parlour environment, noises and the handling. This greatly reduces their stress, makes them much quieter and safer to handle and encourages them to relax so their milk can flow.

Autumn calving began predictably enough in late October 2016. There was the usual fussing, but all seemed to be going along as we had expected. Then the weather changed in early November and we brought the milking cows inside. About fifteen of the autumn calving cows had already calved and all was well, but that quickly changed. We soon started to see grey, bloody diarrhoea coming from some of the calves. Heifer calves were the first to show it, but it rapidly spread throughout the whole group. The calves were becoming rapidly dehydrated, and we spent many hours each day running around with a stomach tube, rehydration fluid and antibiotic. This was so much worse than the first pilot.

The vet came out and swabbed the calf areas and diagnosed cryptosporidium. It's a disease seen mainly in suckling herds of beef cattle, where the young, vulnerable calf is exposed to disease that the cow is resistant to, but a carrier of. The dairy farming practice of taking the calves away from their mum shortly after birth also removes the calf from the source of cryptosporidium infection. In dairy farming therefore, this disease is almost unheard of.

In hindsight it was obvious. When we had done the earlier experiment the barn was brand new, and because it was new there were no microbial pathogens – bacteria and parasites that can cause disease. Over the subsequent four years, things had changed.

We had naively assumed that since the calf was getting protection from disease from its mother's colostrum and milk, naturally all would be well. Colostrum is a substance produced shortly after giving birth, before milk is released. Not only is it highly nutritious, but it contains high levels of antibodies that confer a degree of protection against infection and bacteria from the mum, to the young calf. When they were outside the colostrum and mother's milk had protected the calves just fine, but when they came inside out of the cold weather, they were coming into an environment that was far from natural. It was an environment where bodies were concentrated, and so were any diseases.

The little calves had no natural immunity. A mothers' colostrum might just be good enough to protect her calf, but only if the mum had high levels of antibodies in her milk, and if the calf wasn't exposed to high levels of disease challenge. All it took was one heifer to have relatively low levels of antibody for her calf to become infected and turn that calf into a disease-culturing vessel, pouring infectious diarrhoea into the winter housing.

We urgently needed to find a solution because, by this time, some of our newborn calves had become so unwell that they had died. We upped our hygiene protocols, particularly in the calving areas, and introduced new colostrum management. We realised our colostrum management had been totally inadequate – perhaps some spare colostrum left in a bucket for a day or two or, if frozen, we'd defrost it with some hot water, or leave it out overnight. What this episode taught us was how vital good colostrum was for this new cow-with-calf farming system.

We bought a special colostrum management system from Denmark. This involved collecting three or four litres of good quality colostrum immediately after the first milking of older cows, who tended to have high antibody levels. The colostrum was then frozen and stored in sachets inside a narrow wicker type basket. As each cow calved, we would first ensure the calf was okay, then we'd place a basket of the frozen colostrum inside a machine with a slowly revolving holder, in a bath of warm water maintained at 40 degrees Celsius. Defrosting the precious colostrum gently was essential because if the water bath went over 42 degrees, it would destroy the antibodies.

Next, we would return to the cow and her calf and get the calf suckling. This is essential because the calf needs to learn quickly where its food is coming from – three hundred sucks is the immediate goal. Within an hour we would return with the sachet of newly defrosted colostrum and give the calf a top up from a bucket with a teat or, if in a hurry, using a stomach tube.

Unfortunately, the colostrum management equipment didn't arrive until the middle of December, by which time there were only four cows left to calve. December 2016 was a desperately miserable month as we struggled to keep calves alive. I'd been working with calf forty-six for over a week, re-hydrating her with electrolytes two or three times a day by stomach-tube. She had taken crypto badly, unlike her twin brother who had managed to throw it off and was now doing really well. My hopes for her rose on Christmas Eve when she was lively enough to make it tricky to catch her so that I could give her the medicine. I figured, since she managed to evade me, that I'd got her through the worst of it and she was on the road to recovery. The stomach tube was quite invasive and I didn't treat calves unless they really needed it, so that night I skipped the treatment.

Christmas morning I was on milking duty and went down to the barn at half four to check the calves before milking. There was beautiful calf forty-six, lying flat-out, almost lifeless, in a pool of her own diarrhoea. I slid to the floor, took her wee head and laid it on my lap, and let out a howl of total and utter despair.

Calves were dying one after another. Nothing we did seemed to make much difference in the end. Out of the 55 calves born that winter we lost 17. Another dozen were poorly but limped on for several months before half of them eventually expired too. I was absolutely distraught. We had hardly ever lost a calf in the old way of dairying.

An air of deep gloom hung over the farm team. I'd had enough and started planning to call a halt to this calamitous situation, but I hesitated. The reason for my hesitation was those four youngest calves. They had received the extra, good quality colostrum, and they were all still alive. Not only that, they appeared to be thriving. Was this just coincidence? If we continued with the cow-with-calf system for the spring calving cows, there was a very high risk that the disease would be worse. That's the normal pattern of disease, because by the time spring arrives, but before turn-out onto pasture, pathogen levels indoors hit their peak. I looked closely at those four healthy young calves and decided to continue with the trial.

We moved the autumn calves into their follow-on areas and steam-cleaned the calf creep areas and calving boxes. The spring calving began in early March and we braced ourselves.

We followed a strict hygiene and colostrum protocol ruthlessly. Two, three, five, ten, twenty... fifty cows calved. There was very little sign of crypto. The odd calf that had grey diarrhoea responded well to treatment. It was miraculous. I can honestly say we haven't seen any sign of cryptosporidium in over three years now, but we weren't out of the woods by any means. While our focus had been on dealing with crypto, another disease problem was building.

In the meantime, our herdsman Jim gave his notice. He'd had enough, and I didn't blame him. This was a double blow because his wife Sarah had become our cheesemaker, and she was very good. That previous autumn Sarah had been unwell which led to them both reassessing what they wanted from life, and quite understandably, it wasn't this.

It was September 2017 and here we were again, at a crossroads. Did we go on with cow-with-calf or pack it in? The first year had been a mess. We'd lost so many calves and we seemed to be falling into a bottomless pit of debt with no clear sign of an end to the crises rushing towards us. The decision lay heavy on my shoulders, but it didn't just affect me, it impacted the whole farm team. There was Barry the stockman, Abe who worked part-time, and Erinna, Jim and Sarah's nineteen-year-old daughter, who was doing her agricultural training on the farm. I asked them, 'What now? Do we go on?' With finances on a knife-edge I couldn't afford to hire another herdsman. Erinna and I would need to share milking duties, with Barry and Abe in support. It would be hard work, but they were all keen to give it another year.

The second year was transformational. There was little sign of crypto and the cows were so much more relaxed, even letting milk down in the parlour which meant the butterfat levels in the milk were improving. We were settling into some kind of routine. With the calves being much healthier, the stress in their mothers was noticeably less too. Unquestionably when a calf got ill its mother knew immediately. It was often her who tipped us off that there was a problem; she might be bawling or acting in an unusual way. When you're working closely with these great animals, their body language is pretty clear.

There was this other niggling health problem rearing its head. Muggy calm weather seemed to trigger it. The calves would show a fever and if you missed it, within 24 hours they'd be un-saveable. Vet swabbing revealed a type of pasteurella pneumonia, and there was no vaccine. Early detection followed by a shot of antibiotic and anti-inflammatory sorted it, but with up to a third of the calves requiring treatment, this was unsustainable.

Interestingly, the intensive swabbing and pathogen investigations had detected a low level of a nasty disease that we had never seen before – salmonella. Our vet speculated that the salmonella might be acting as a trigger for the pasteurella pneumonia. We could vaccinate against salmonella, but all females over six months would need to be done, and it wasn't cheap. It seemed like a bit of a long shot and I delayed making a decision because we had our hands full just getting through each day.

We knew from our organic conversion experience that it would take years before the cow-with-calf system would be firing on all cylinders. What we didn't know in that first year was just how bad it was going to get before we managed to pull out of the steep nosedive into the financial black hole. And even then, if we got in too deep, would we ever be able to pull ourselves back out?

Chapter Twenty Six

Wilma

With the trial progressing we needed to make a decision about our branding. The key to making this whole system transformation financially viable – and genuinely circular – was using all the milk from the herd to make good quality cheese on the farm. To do that we needed to sell cheese in significant volume, and to do that meant we needed a strong brand.

David and I agree with our critics that calling ourselves 'The Ethical Dairy' was arrogant. It took us a long, long time to agree that our brand name would be The Ethical Dairy, and we resisted it every step of the way.

We had originally launched our cheese range under the brand of 'Finlay's Farmhouse Cheese', nothing controversial there. Once we re-started the cow-with-calf experiment in 2016, David and I spent months trying to decide on a brand name that would reflect our dairying methods. Our favourite was Nature and Nurture, but we had over twenty name ideas, including Lucky Cow, Cow Pleaser and Pastures New. We did a survey with our team, trusted customers and friends, but there was no clear winner. In fact, for most of the name ideas people thought we were either being too preachy or gimmicky.

The one area of business that neither David nor I are good at is marketing. People may scoff when we say that, but what they don't know is that we've worked with the extremely effective and astute Lorna Young who guides us through this quagmire. When I say 'guides' I should maybe say directs or encourages, but sometimes she just bosses us. She had been persistently nagging us for at least a year saying that we were the Ethical Dairy. And not just an Ethical Dairy, but The Ethical Dairy, as if there couldn't be another ethical dairy.

Initially David and I just ignored her. There was no way either of us would ever stand up in a meeting and introduce ourselves as coming from 'The Ethical Dairy', and so we tried to move on, but Lorna was playing a long game and she wasn't going to lose this one.

Our dilemma over the whole naming thing was that we wanted to find a name that would immediately tell the customer that we keep the cows and calves together, and that in everything that we do, we try to do the best for the environment, animal welfare and society. How about Decent Dairy? She shook her head, and by this time she had back-up.

Our graphic designer, Ian Findlay, who is also our photographer, had started working on an idea. Out of the blue he presented us with a logo using The Ethical Dairy. It looked good. We bristled. We reluctantly agreed that the cheese packaging design work could start with that as a 'working' brand name, with David and I secretly planning to change the name further down the track once we had triumphantly come up with a better idea. Eventually in the winter of 2017 we caved in; relenting in part because we noticed visitors on the farm tours offered by Cream o' Galloway were, unprompted, starting to use the word 'ethical' too.

I want to say that branding ourselves as The Ethical Dairy is the best decision we ever made, and from the point of view of the business that is definitely true. At public events, people come to our stand and ask what makes us ethical – the perfect conversation starter. In delis, customers ask for the ethical cheese – the one where the cow keeps her calf. The branding works and we are both now very comfortable with the name, but there has been a personal cost.

While having a loving, encouraging online community is truly uplifting, losing friends in your neighbouring farming community is very painful. Being in an online bubble isn't good for us, we need to get back out there and laugh with neighbours as they tease us, even if they are teasing us for our apparent arrogance.

We launched the brand with a press day in March 2018, and our local ITV news, local BBC radio and local press all attended. It was the usual last minute gathering of banners and props, and we have lots of photos and footage from the day that we still use. The event was hugely successful and launched the new brand name, and our claims about this radical new approach to dairy farming, with a fanfare. The behind the scenes story of that day was, for me, quite different.

I was just out of hospital having had my second mastectomy after precancerous cells had been found during one of my, then regular, scans. I was offered an early date following a cancellation and decided to take it even though it could potentially have meant I would miss the pre-arranged press day. The close team around me obviously knew, but the journalists didn't.

Anyone who has had a major op and been told not to lift anything will understand the frustration. I spent the whole day trying not to appear as a prima donna, standing to one side doing nothing as bales of straw were being moved around for different settings. I was trying to avoid being in a situation where someone would pass me a heavy prop to hold for a minute. Most of the photos show David holding a 15kg cheddar while I had a cheeseboard selection on my lap. There is one photo of me holding a 5kg Rainton Tomme cheese. I probably shouldn't have done that. Maybe that's why I ended up with a painful seroma a week or so later.

Chapter Twenty Seven

David

The rest of the dairy industry had already written us off as, at best, a marketing ploy and at worst a joke, but the launch of The Ethical Dairy brand was met with the occasional raised eyebrow and a raised hackle. A visiting group of Danish dairy farmers made the comment, 'So, if you are The Ethical Dairy, what does that make us?' which pretty well summed up the general reaction. My cousin in London, an ex-city trader commented, 'Sounds a bit pompous, doesn't it?' Other than that, the industry figured the brand was just a sales gimmick and ignored us.

What it did do was get the public to ask, 'So what makes you so ethical?' giving us the opportunity to explain what we did and to invite them to visit the farm to see for themselves. The launch of the brand was a very public statement of opposition to the mainstream trajectory towards intensification, and a statement of intent that we were determined to find a different path. A 'kinder' path is what we claimed.

'What we're doing is de-intensifying dairy farming,' we said, 'it's the opposite of what's happening elsewhere in the industry.' It was strong stuff.

'By branding ourselves The Ethical Dairy we are not demonising farmers, nor claiming to be perfect. What we're trying to convey is that in everything we do we aim to meet the highest ethical standards.'

We continued, 'If all goes to plan we hope to demonstrate that food from the dairy industry can be produced with compassion for our animals, for our people and for our environment. We also hope to show that far from being expensive, food produced this way can actually cost us less. This is just the beginning.'

And so, indeed, it was. Big statements, bold claims and still so much to do.

As the cows and calves went out to grass in the spring of 2018, we were full of optimism for the cow-with-calf experiment. At last, we were getting on top of the system's biggest problems. Sure, if this was ever going to be a food production model that could be profitable there was still lots of work to do, but at least we could now see the potential.

I know it's difficult to believe, but looking back at independent financial data for our later organic years showed that we could produce milk at the same cost per litre as our intensive colleagues. The problem has always been the cost of getting relatively small amounts of milk transported, processed and placed in front of the customer. I also realise that people will look at the price of our cheeses and note that they are much higher than supermarket cheeses. There are two reasons for that. Firstly, we've just gone through a hugely expensive experimental exercise and borrowed a lot of money that needs to be paid back, with interest. Secondly, we are a tiny dairy producing artisan cheese using natural processes and traditional, labour-intensive techniques. This type of cheese is, by its very nature, more expensive to make than factory-produced cheese.

My dream is that a milk processor sets up to process cow-with-calf dairy milk and delivers it to the general public. It would require quite a few farmers to be following this model of dairying within a fairly small geographic area, but if that could be done, then the cost of processing and selling cow-with-calf milk would plummet, and it would make an important point.

It has always been the position of the industry that you can't have plentiful, healthy and affordable food, and at the same time have high social, environmental and ethical standards. I used to believe this as well, but not anymore. You can have it all.

What this project has proven to me is that we can have a satisfying, balanced lifestyle career, contributing positively to the global supply of healthy, affordable food from happy, contented animals grazing diverse pastures produced by healthy soils with no toxic side effects. And we can do it profitably, which is so important. Over the years I've been approached many times by people offering to sponsor retired cows, and that's very generous of them. If we were farming this way simply to indulge our own interests, or if we were transitioning to become an animal sanctuary, then that approach would be appropriate. But that's not what we are doing.

Ultimately we want to see cow-with-calf dairy farming become mainstream. When we launched The Ethical Dairy brand we announced that we wanted to 'revolutionise' dairy farming. That remains the case. To do that means we need to demonstrate that this way of producing nutritious food can not only be viable, but that it can be profitable. It's numbers, not rhetoric or values, that will convince a reluctant dairy industry – and governments chasing easy options – to start walking a non-intensive path.

The numbers are now starting to make the argument for us. Every year we used to go through more than 100 tonnes of soluble fertilisers; 75 tonnes of lime; several thousand litres of weedkiller; hundreds of doses of worm drench; hundreds of doses of antibiotic and vaccines and 240 tonnes of purchased and mineralised, cereal and soya stock feed.

We now use no soluble fertilisers, only 20 tonnes of lime, no weedkiller and only a handful of doses of worm drench, about 25 doses of antibiotic each year, one course of vaccine which ends this year, and 100 tonnes of purchased feed which is just dried lucerne – a legume. No cereals, no soya and no minerals.

Right now, the prices for the inputs that we no longer need are shooting up. In total, what we've cut out of our system would, this year (2022), cost more than £100,000. So, this way of farming directly reduces our costs. What will be surprising to many, is that our productivity is pretty good too.

We are rearing an average of one calf per cow in the herd. The calves are growing at more than twice the rate of the calves when we were bucket rearing them. Since we are no longer bucket feeding calves we need less staff, so our staffing requirements have also reduced. We are re-investing that staff time in looking after the herd's welfare, but it could be done with less labour.

The cows are milking really well. On this 100% forage diet a cow is normally expected to produce around 4-4,500 litres of milk in a year. We know from our trial work that our calves drink an average of 15 litres of milk a day. Over the five to six months they are with their mothers they are drinking 2-2,500 litres of milk. We are getting an average of 3,000 litres per cow, and every year that gets better. So our cows are each producing an average of 5-5,500 litres a year and that is still increasing. Without a doubt, the suckling and the low stress environment stimulate milk production.

With the calves growing so quickly, they are reaching marketable and breeding ages more quickly too. So, instead of 100 milking cows with all their youngstock through to 26-32 months of age, we now have 120 cows and youngstock to 8-26 months of age. This means the farm is much more resource efficient than it was before, and it's also more resilient.

It doesn't matter to our farm that the price of fertiliser is going through the roof. That doesn't affect us at all. Even feed price rises only affect us a little bit. From an economic point of view, what this system has done is enabled us to throw off the shackles of corporate parasites who have created an illusion in agriculture of a need for their toxic products, as they suck the life out of farmers and the planet. Maybe the biggest change in this farming system is that we have taken back control.

When you consider the benefits to biodiversity of not using all these chemicals, the lack of run-off into waterways and agriculture's contribution to antibiotic resistance; the benefits of not using food crops grown on land where tropical forests once grew, or feeding our cows cereals that could be eaten by humans, the question that should really be asked is why isn't everybody farming this way?

I expect that within the next two years our unit cost of producing a litre of milk – in an organic, cow-with-calf, 100% forage system – will be no more than our 2000-cow neighbour. Looking at the way the market is going, we might reach cost of production parity with our intensive neighbours well before then.

It's not the cost of milk production in this cow-with-calf system that's expensive; it's what's involved in getting our small scale artisan products to our customers that adds the cost. If the UK dairy industry switched to this system across the board, I am confident the price of milk in shops would be not much different than it is right now. Will the industry switch? Well, that depends on what the public and the government demands.

Chapter Twenty Eight

Wilma

The launch of our brand name in 2018 was just the beginning of a year that was to prove transformational. By this time we had been joined by an experienced artisan cheesemaker and our confidence in the quality of our cheese was soaring. The big product launch to the trade was planned for the Speciality Food Show at Olympia in London, which started on the first weekend in September. I'd done my fair share of trade shows over the years and knew that this is one of the most important food industry events in the country. Shows like Speciality are an opportunity to meet with buyers from across the food industry – from independent delis to large retail multiples; the perfect place to launch an exciting new cheese range.

Then we heard there was a consumer event, Go Organic, taking place in Battersea Park the following weekend. Why not do both? If only we'd stopped there. We had four days in between the shows so we figured we could host a pop-up event in London to try to catch the interest of national press and cheese buyers. Naturally, in April, a pop-up in September seemed like a great idea. There was one other thing we decided to add into that intense ten day period.

We had been thinking about crowdfunding for years. As part of trying to figure out a financial solution to make this farming transformation viable, we had explored the idea of setting up a

Community Interest Company and offering shares in our business. But, at the end of the day, we knew ourselves well enough to understand that we needed the freedom to make changes and decisions very quickly. We had already experienced the suspension of our first cow-with-calf trial. What if something like that happened again and the majority of shareholders had only invested in the project because of the cow-with-calf angle? David and I may think of ourselves as having integrity, but that might not be how others would perceive us if we were forced to call a halt to the whole thing once again.

The one big financial investment that remained to make this farming transformation work was to turn Andre's vision for a cheese production space and cheese store into a reality. At this stage we were still making cheese in the ice-cream factory. In the summer we made ice-cream four days a week, then two days of cheese, and during the winter months that switched around. The footprint of a cheese vat is much bigger than that of an ice-cream freezer, so we could only fit in a 300 litre cheese vat. What this meant was that we couldn't make very much cheese at all, and we certainly didn't have the production capacity to use all of the milk produced by the farm.

Since we were making cheese in very small batches it also meant our cost of production was unaffordable. The reason is that it takes one person exactly the same amount of time to make a small batch of cheese, as it takes them to make a batch ten times as large. The main cost of traditional cheese production is in the staff time, not in the price of the milk. We urgently needed to get that cheese dairy up and running, which meant we needed to raise the money to do it.

We had applied for a European grant, the purpose of which was to help food producers invest in manufacturing equipment. That grant had been approved, but we still had a big shortfall. The Go

Organic festival seemed like the perfect opportunity to launch a campaign, so we revisited the idea of crowdfunding. We got fliers printed to hand out along with the samples of cheese. It was a lot of planning to get done in the next four months, but we were confident it could be done. We left the launch of the crowdfunder in Lorna's hands on the Saturday of that busy weekend while we focused on the events.

The Speciality Food Show and Go Organic were both exhilarating, and of course the opening question was nearly always 'What makes you ethical?' David and I quickly developed a routine whereby he answered questions while I took contact details and offered samples.

David's voice was giving up and I tried to get him to find a more concise way of answering all the questions, but no, he was on a roll. One person we both remember vividly approached our stand at the end of day three. She was a fellow stand-holder, a fiery American redhead who was quick to tell us that she had organised an animal rights demonstration earlier in the summer. 'David, you have met your match' I thought, but no, the two of them ended up shaking hands and hugging. She was a realist. There would always be people who ate meat and dairy, she said, and if we could set a new standard for animal welfare then she wished us well. We've bumped into each other a couple of times since, and the mutual respect continues.

The pop-up event was the big unknown of that trip. We have a good relationship with Chris and Denise Walton of Peelham Farm who were also going to be attending the Go Organic festival. They already had a market for their charcuterie in some London outlets so we decided that hosting a joint dinner promoting their charcuterie and mature beef, as well as our cheese, could be a draw that would benefit us both. But the question was where and when? It quickly became obvious that no restaurant was going to give up a

Thursday or Friday night's income to accommodate a pop-up, especially for people they hadn't even heard of. Our contacts just dried up.

Then along came David Cairns, who had worked with Chris and Denise off and on for a few years. If ever there was a case of 'it's not what you know, it's who', then this was it. David is a food activist working in Glasgow but with lots of contacts in London. He had worked with Leila McAlister of Leila's Shop in Shoreditch on supporting the homeless. Leila's shop was closed on a Monday, and she said we could have it. So the evening after Go Organic we would be running a pop-up restaurant in Shoreditch, when any sane person would have returned home to relax and recover after taking part in two major events.

For a control freak like me the arrangements around the pop-up were uncomfortably loose. We had David Cairns assuring us that everything would be fine, but with nothing like the detail we felt we needed. Leila's normal chef was on holiday that week and her guest chef was James Ferguson who had exactly the experience in 'Nose to Tail' cooking that we needed. Denise and James managed the occasional phone call to discuss the menu, but it wasn't until the Friday, just three days before the pop-up, that we all met. He was blown away by the ingredients.

Leila's café is small, just 24 covers. How many people do you invite for 24 covers? 30? 40? 50? We decided to go for 50 and hit it more or less bang on, welcoming a great combination of journalists, bloggers, politicians, chefs and cheesemongers.

The most memorable part of the whole ten days for me was the food James produced. He was even generous enough to share his cooking methods with us all. It was way past midnight by the time we got back to our AirBnB, and we knew we wanted to leave early the following morning to miss the traffic, but could we get to sleep? Not a chance. There was just too much adrenaline coursing through our bodies. In that ten day period we had attended two major foodie events, hosted a pop-up meal in Shoreditch and launched a crowdfunding campaign. And there was one other thing, something we could never have planned for, that happened that evening, the impact of which we still feel today.

Chapter Twenty Nine

David

The London fortnight was an exhilarating whirl and we were high on adrenaline for much of it. It was only as we were sitting with our friends Chris and Denise having a relaxing drink that we began to realise the enormity of what had been happening in the digital world. That weekend our project on the crowdfunding platform was 'trending'.

Before that, however, I took a phone call and discovered the BBC Scotland Dispatches exposé called *The Dark Side of Dairy*, something I had been interviewed for some months previously, was about to be broadcast.

I'm not sure how the BBC had found us. We'd done some stuff for them in the past, with mixed feelings at our end. These things are always a bit of a risk as you have absolutely no input into the editing or end product; how you come across – and consequently how you're perceived by viewers – depends very much on the agenda of the production team. And so it was in this case.

A documentary investigation into the live export of young calves out of Scotland turned into something of a witch hunt. We were being used as a stick to beat the dairy industry with, and the dairy industry was not impressed.

To be honest, I didn't really care. I find the 'ship 'em or shoot 'em' mentality towards these young dairy bull calves totally abhorrent and I know I'm not the only one. It's a consequence of the incessant drive towards greater production of milk and the breeding of extreme types of dairy cow where the bull calf is seen as a valueless, waste product, rather than as a sentient being deserving of respect.

As soon as the programme was broadcast, the exposé hit the proverbial fan and social media went crazy. We were being attacked from both sides with a torrent of abuse from both vegan and farming fundamentalists, but the support from the public was fantastic.

Our part in this exposé meant we were toxic as far as the dairy farming world was concerned. I always measure how much I have stepped over the 'acceptable' farming society line by my bellwether cousin's reaction. It took dear Kate, his wife, a full two months before she could get us to sit down to a rather strained dinner together.

While it is okay for farmers to criticise others within our farming circles, to be associated with a programme that publicly shamed the entire industry was absolutely inexcusable.

Chapter Thirty

Wilma

With all the planning for the London events, and the co-ordination of the crowdfunding campaign with that trip, it hadn't crossed our minds that the BBC programme might be broadcast that same weekend. As it turned out, the programme was screened on Monday 10th September, while we were hosting the pop-up meal. We got a hint of what might be coming that morning, when a short clip of David, choked with emotion, featured in a news item on the BBC promoting the programme. The clip was also used in an online news story, and it began to be shared widely across social media. We had no option but to switch off our phones and try to forget about it.

On the drive home, in the early hours of the next morning after the pop-up meal, we tried to watch the programme online but our mobile broadband just wasn't up to it. Social media gave us the gist of how it was being received. Looking back now, it's difficult to describe how that time felt. It was utterly overwhelming. Lorna had been in touch to say that feedback was very polarised. She had been managing social media and responding to requests for statements that were flooding in from the press, and even from the Scottish Government. The public was very supportive, the press had lots of questions, but vegan fundamentalists and other farmers, many of them local to us, were on the attack.

The week that followed was a rollercoaster. Our approach was praised by MSPs in debates in the Scottish Parliament and lauded by complete strangers, but we were snubbed by farming friends – and it didn't stop with us. Members of staff were left feeling uncomfortable within their own social circles as local people – many of them connected to the farming community – made it clear they thought David should not have taken part. There is no doubt in my mind that appearing in this programme was good for our business, but that wasn't why we did it.

We took part because we wanted to show there is another way to do dairy farming; a way that means bull calves don't need to be shot at birth, or exported at a tragically young age to endure short lives spent in intolerable conditions. Our system was starting to work. We knew there were solutions to all of these industry dilemmas and so we felt an obligation to talk openly about what we had discovered. But we were lambasted. We tried to develop a thick skin to survive the fallout as many dairy farmers in Scotland looked for someone to blame for the highlighting of widespread bad practice in the industry.

All this media attention coinciding with our crowdfunding campaign was sheer coincidence, but that didn't stop accusations that we were exploiting negative publicity for the dairy industry for perceived personal gain.

A crowdfunding campaign is a major risk. Getting it right is so important and we had procrastinated about whether or not to do it for years. There are so many options to consider, including how much money we wanted to raise. What we were trying to fund was the cheesemaking infrastructure – essentially the building renovations and manufacturing equipment – that would allow us to produce cheese in volume. This was a critical step in enabling all of the milk produced by the farm to be used on-site. And making use of all the milk ourselves was the key step in creating the revenue stability that would allow this entire farming transformation to work financially. It wasn't a straightforward ask.

We really needed £600k. We had raised a third from the European grant, but we had to figure out where the remaining £400k was going to come from. We had explored so many options over the previous few years, including so called 'ethical investors' who were interested, but only as a loan, and even then the interest rate they wanted was eye watering.

Setting £400k as a crowdfunding target was utterly ridiculous. So, the decision was made to set a target of £50k, which in itself was ambitious. Ultimately the goal of crowdfunding was to raise enough money to demonstrate to our bank that we had sufficient interest from members of the public to justify the bank lending us the remainder. The campaign wasn't just about raising money, it was a way to demonstrate public support for our approach, in order to prove we were a 'safe risk' for the bank. We were on a knife edge of viability.

One thing we did understand was that as soon as you go live with a crowdfunding campaign you will be full-on for the duration of it. We knew we would have to be prepared. We would need blogs, photos and video footage to promote our campaign on social media and to answer all the questions we could foresee. Ian, our designer and photographer, had captured beautiful video footage and photos of the cows and calves the previous autumn, and we wanted to use that footage to tell our story.

There was also a healthy dose of luck involved. We chose the Crowdfunder platform and we cannot thank them enough for their support. Within a day of the campaign launching we got a message from an advisor who'd been assigned to our project wishing us good luck and offering support and advice. He also mentioned that he knew a relative of David's. We both assumed the relative would be David's cousin in London who is a well-connected retired stockbroker, but no. It was David's nephew who was in his early twenties, and our advisor was his former university flatmate. When you get to your sixties, this kind of wakeup call that your network is no longer 'where it's at' is humbling.

I don't think David and I ever really expected the crowdfunding campaign to work. We had been down so many blind alleys trying to raise funds that we questioned why this should be any different. The idea of 'trending' on Crowdfunder had never occurred to us. We checked progress several times a day. The donations kept coming. We got immense pleasure when we were told that several neighbouring farmers also started their day by checking how much money we had raised. Is there a word for the opposite of Schadenfreude? Probably smugness. Admittedly, there was a little bit of that.

The realisation that so many people really wanted this to succeed took our breath away. And they didn't want just us to be doing it; they wanted farms up and down the country to be providing local communities with dairy products from cow-with-calf dairy farms. We were exhilarated and heartened by the amazing support. We hadn't anticipated how much hope our system was giving to so many people. The demand for kinder, caring farming was repeated over and over again. The need for farming to change was so clear, the public wanted us to help make that change happen.

One of the most surprising things was the number of vegans who contributed, and I do mean vegans, not 'near-vegans' who couldn't give up cheese. These were people whose partners or children weren't vegan and if there was going to be cheese in their fridge they wanted it to be from The Ethical Dairy. There are only a few times in the past thirty years that David and I have said 'This is it, we've done it'. The crowdfunding campaign was one of those times.

In the end almost 600 people donated and every single comment they left touched us. Some were extremely generous and some left messages along the lines of 'I'd like to give more, but can only afford £5 this month'. Every single donation, every share on social media and every heartfelt message meant so much to us.

We hit our target of £50k a week before the end of the deadline and by then we had started to look at the fine print of the platform. We were advised to set a 'stretch goal', so we spent a crazy Monday night debating ideas about what new project we could bolt on to keep the momentum going. In the description of our campaign we had mentioned that we wanted to make what we'd learned about cow-with-calf dairy farming available to others – we called it open source farming – so we had a think about what might help us achieve that. On the spur of the moment we pledged to organise a 'small' conference on our farm. The idea of the Ethical Farming Conference was born.

Chapter Thirty One

David

Back home, after returning from London, was like waking from a deep dream as I came to terms with the realisation that I had crossed a red line in the dairy farming community. Memories are long and forgiveness short. The guys on the farm advised me to stay well away from social media. I haven't been to a local livestock market or dairy farming event since, but the flood of appreciation from the public opened a door to a new, deeply supportive community.

The stark reality of life on an intensive dairy farm, and the wrongness of it, was brought home to me when our former stockman returned to the farm. Charles was always a keen proponent of the idea of cow-with-calf dairying, but his enthusiasm for it probably became jaundiced by our chaotic experience of cow-with-calf in practice during our 2012 pilot.

After working with us for ten years he had left to take on the challenge, and the big money, of an assistant manager role at a mega-dairy. 2,000 cows, milking three times a day and with 365 days a year housing. That experience had taught him a lot about the mess and misery so often found in the wake of the industrial farming model. He had seen just about everything that can be seen on a dairy farm. The hours were long and the pressure relentless. Twelve hour days, twelve days in a row, then just two days off – and if someone else was off work through ill health, the hours and days became even longer for the rest of the team. Charles found he hardly saw his young family. When Jim left, Charles was on the phone enquiring after the job. He conceded there was more to life than money, but I was in no position to employ him at that point as our finances were precarious. I told him to get back in touch the following year, which he duly did.

Charles was a changed man. Retaining staff on large dairy farms is a major issue and for Charles the pressure of the mega-dairy had triggered anxiety and depression. He isn't alone. Taking him on was a risk, but I knew he believed in what we were trying to do, he had excellent training, practical experience and he also had something to prove. I offered him the job. I wasn't going to find anyone better to take on what the industry considered to be a no-hope system. Things were a bit rough to start with as he tried to adjust from the hectic pace of the mega-dairy system to our slow-paced, low-stress one.

Without doubt, the system on our farm now belongs to Charles. He has embraced full responsibility for taking the model forward to become a compassionate, resilient and profitable dairying system. It was Charles who pushed to vaccinate the herd against salmonella, which we did in the autumn of 2020. The impact was immediate. Pneumonia cases in the calves fell to just a handful that winter, and we continued the vaccination programme into the following year with similar success.

Before Charles returned, I had looked for a natural solution. We had met a chap called Aled Davies from Wales, who had been head of a European sales force for a leading agri-chemical company. Aled had discovered he was intolerant of antibiotic and reckoned that if he ever got a serious infection, it could be game over for him. He went on a Nuffield Farming Scholarship to research alternative treatments to the use of antibiotic, and he then set up a company supplying these alternative therapies.

Our interest lay in a couple of his products. These were probiotic solutions that we would either mist round the cattle barn or add to the water supply. The thinking was that we would swamp the microbes' environments with these probiotic bacteria in order to out-compete the pathogenic bugs. Apparently, we were told, the pig and poultry industries were already using this system to reduce their dependency on antibiotic use, and Aled explained that some hospitals in Canada were even using them to control the antibiotic resistant bacteria MRSA and C-Difficile. Using nature to control nature. It wasn't cheap but it was intriguing and the idea made sense to me.

We used the misting version of this system over the 2018/19 winter and it certainly seemed to have some impact on the pneumonia which improved, with around a third less cases. We also used the water version, adding probiotics to our water settling tanks, these were fed by the farm reservoir and they supplied water to all the cattle barns and stock fields. Aled had assured us this system would keep our water pipes and troughs clear of organic debris, and that it would also deal with the micro-film that covered all under-water surfaces; a film that can be a platform for pathogens to re-infect after cleaning and disinfection. Of course, Aled was a salesman, so we only half believed all this, but he was spot on, certainly, as far as the water bugs were concerned.

This experiment with natural approaches to disease management was, in my mind, simply an extension of the natural systems-based approach we had adopted on the farm. By harnessing natural systems we were finding the health and productivity of the pastures and animals was probably even better than it had been previously. Some fascinating science coming from a few cutting-edge soil scientists might explain what lay behind this, but I'll try to give the farmer's version of the science.

Now that we weren't spraying weedkiller, the pastures – instead of being just ryegrass with a little bit of clover – were filling with herbs, clovers and many different wild grass species. It is the diversity above ground that creates diversity below ground. As this diversity flourishes, so does the health of the soils. Healthy soils contain many millions of microbes. These microbes are able to access minerals in the soil that are unavailable to the plant roots. Although the microbes can't produce their own food, the plants can.

Above ground, in the plant leaves, are millions of tiny factories that can make food (carbohydrates) out of carbon dioxide and water using the energy of the sun. Food that's surplus to the needs of the plants moves down to the very fine root hairs of the plant where some kind of exchange goes on. Plant produced food is traded for microbe mined minerals.

When we were applying soluble fertilisers and weedkillers to our pastures, we were disrupting that soil lifecycle and creating a dependency on feeding the grass from a fertiliser bag. That was bad enough, but much of those fertilisers and pesticides no doubt got washed into our water systems and ultimately into our seas where they would have done untold damage to ocean wildlife.

An important function of a healthy soil is that it can capture carbon. What seems to happen is that microbes munch microbes, and some even munch microbe poo. This process goes on until there is a substance left that nobody wants to munch. In a healthy soil, so long as you don't plough and cultivate it, this starts to accumulate. This substance is a major component of soil organic matter, and soil organic matter contains large amounts of carbon. So, as the soil health grows, so does the soil organic matter and, therefore, the amount of carbon contained within the soil increases. This is what is known as soil carbon sequestration.

When I had returned home from my consultancy work all those

years ago, I had been uncomfortable with the methods used by fertiliser sales reps when they took soil samples. I'd also been a bit suspicious about the accuracy of their soil analyses; after all, they were doing soil sampling for 'free' and I knew there were no free lunches in the agriculture industry's supply chain. Part of my farming consultancy work had been taking soil samples, which meant I had been trained how to do it professionally. So when I returned home to Rainton, and for the next 25 years to come, I had been regularly sampling the soil on the farm myself and sending those samples to an independent soil laboratory for analysis. I did this initially just to determine soil acidity, and also the phosphate and potash requirements, but more recently I had paid for soil organic matters data as well.

I was very curious to know how our soil organic matters had changed over that 25-year period as we transitioned from conventional farming to organic and regenerative methods. I knew that this lab had been doing soil organic matters routinely back when I was doing consultancy work, and with a bit of digging I discovered that this had continued right up until 2013. I had started paying for organic matter assessment from 2016, so realised that if I could find that old data, then I would be able to track how the organic matter had changed on our farm over the years. It took nine months and a bit of inside help, because the data was stored on an old computer system that required specialist knowledge to access, but we managed to retrieve the data. It felt like we'd uncovered buried treasure.

Once we'd unscrambled the data it showed that over the whole farm, and from several thousand different sample points over the 25-year period, the soil organic matter had increased from 11% to almost 14%. Unfortunately, there is no straightforward method of translating this into tonnes of carbon stored per acre. First off, the sampling is of the top ten to twelve inches of soil. What about the next three feet of soil? Some recent research indicates there is as much carbon in the next yard of soil as in the top ten inches. Then there's the phenomenon known as 'carbon migration'. That is, carbon in the top ten inches increases faster than in the next yard. Finally, there is the change in 'bulk density'. That is, as the organic matter in the soil increases, the soil gets lighter in weight which affects the carbon percentage calculation.

Of course, we had no idea how much these variables had changed; we are a working farm not a laboratory. I discussed this with a soil scientist who reckoned that if we deducted one third from the calculated figure, this should more than cover these uncertainties. Having done that, the raw figure of three percentage points increase translated into six tonnes per acre of sequestered carbon per year over the last twenty years. Less a third gave us four tonnes of carbon sequestered. Now, we'd been getting carbon emissions audits every two or three years and our emissions were assessed to be about 1.8 tonnes per acre per year. To be on the very safe side, we could halve the four tonnes sequestered and still claim to be 'net zero' – and this didn't even include the carbon sequestered by the 35,000 native trees we'd planted twenty years ago!

Another fascinating thing about what was going on in our soils and herbage was the nutrient concentrations. For as long as I could remember, the farm stock had always suffered from copper, cobalt and selenium deficiencies, and so we had to feed them mineral supplements. Now, instead of fertilised ryegrass, we had a diversity of plants, many of which were deep rooted and high in mineral nutrients. This becomes visible in mid-May when the fields turn gold. Visitors ask what crop we're growing. Dandelions! We wondered what would happen if we stopped feeding the animals mineral supplements. Would the cows get ill? Would they fail to breed? Well, we stopped feeding vitamins and minerals to all the stock more than three years ago. We haven't noticed any difference, except that when we wean the calves from their mothers, we need to give them a slow-release copper bolus, a supplement, to get them through that transition.

A significant benefit of this ecological farming is that we use a lot less lime to keep the soil acidity balance right. In the olden days we needed to apply around 150 tonnes of lime to the fields every couple of years – we haven't needed to put any lime on for the past six years. The prevalence of mushroom 'fairy rings' in our fields in recent years is also quite remarkable. These rings can be just a foot or two in diameter or they can be fifteen or twenty feet wide. Some fields have one or two rings, while others are covered in them. From August to October, I was supplementing my meals with a feast of freshly picked brambles (blackberries) and button mushrooms. I asked a soil ecologist about this, and she confirmed these fairy rings were a good indicator of healthy soils.

A book that is well worth a read is *10% Human* by Alanna Collen. In it she claims scientists have found that of all the living cells in our bodies, only ten percent are human. Most of the non-human cells are found in our digestive tract and they are essential to our health and wellbeing. They are much tinier cells than our human cells, but they form an ecology that should be given careful consideration.

The microbial biology in our digestive tract needs constant stimulation as does our immune system, because what happens when we have an immunological army but no enemy? It becomes unfit for purpose or turns on its own. Much of the food we eat is over processed and utterly lifeless, resulting in the loss of a healthy, biodiverse, gut ecosystem and the stimulating effect that has on our immune system. Perhaps this sterile food is why there's been such an increase in diet related disease, allergies and auto-immune issues?

Another very interesting but very technical book is *The Probiotic Planet* by Jamie Lorimer. In it he describes how the unseen, parallel universe of the microbe mirrors the one we can see. The story of how re-introducing wolves – a keystone species – to Yellowstone National Park in America changed the entire ecology of the park. In a way, we see the same thing with the badger here on the farm.

Lorimer describes how there are similar forces at work at the micro level about which we know very little. His example is how the presence of the hookworm, a creature that co-evolved with humans, in a human digestive system can totally transform the gut ecology such that certain digestive diseases could be corrected. In a similar way we saw how controlling a minor pathogen like salmonella had a huge impact on the susceptibility of the young calves to pasteurella pneumonia.

Books like these have strengthened my desire to minimise the number of disruptive interventions we make to our ecological farming system, because we just don't know what the consequences of those interventions might be. At the same time, Wilma and I are determined to provide food to our customers that are more than just sterile fat and protein. Our raw milk cheeses are a living ecosystem, devoid of pathogens, for which we test extensively, that bring beneficial gut ecosystem stimulation from their probiotic contents. Healthy soils produce healthy crops and healthy animals and, from them, healthy food, but it goes further than this.

The growing evidence from the world of microbes is that we are a part of that microbial world, not apart from it. We are dependent on the microbes and they in turn are dependent on us. Careless action that disrupts the microbial cycle is bad for them, bad for us and, most likely, bad for the wider environment. I therefore cannot understand how individuals, with the best of climate intentions I'm sure, can propose that we further disconnect ourselves from these complex but essential cycles of nature by eradicating our world's nutritional dependence on livestock.

Over the past 25 years of transforming our farming system the overriding lesson I have learned is that nature knows best. So, recent claims that we should grow our food in giant fermentation silos while rewilding the landscape, go against everything that nature has taught me. I cannot help but conclude that the industrial manufacture of proteins, currently being proposed as a climate solution, is going to be bad for our health and wellbeing by disconnecting us at a species level from that vital microbial dance.

Chapter Thirty Two

Wilma

An unexpected advantage of the crowdfunding campaign was the relationship it built with our customers and with those who follow us on social media. The depth of that connection surprised us. We had been used to engaging with members of the public – of course we had, we'd run a visitor centre for over twenty years now – but this was different. The people who connected with us through The Ethical Dairy brand were doing so largely because they shared our values, and that is a very different kind of connection.

Lorna has been our marketing guru throughout and I can remember when she first said we needed to make David and the farm more visible within the brand. Oh how we laughed at the very thought of people being interested in the minutiae of day-to-day farming and of David's latest insights! But David started reading the emails and social media comments. Our supporters were people he could really relate to, and he genuinely wanted to answer their questions and to find out what they thought of our system. Slowly but surely David became an unlikely blogger. These blogs give our customers the reassurance that we are what we say on the tin, and the comments those blogs generate give us a great deal of food for thought.

At this point cheese sales were mainly to our local customers and a couple of Scottish cheesemongers, but at Christmas the online sales really escalated. We had launched a website with e-commerce, and our ultimate goal was for 20% of our sales to be online. Initially it was just a trickle, mainly crowdfunding supporters who had selected a cheese subscription as their reward, but word of mouth and social media have been our friends. Word spread and many more people started to buy our cheese regularly; but nothing compares to Christmas. Throughout our short life as cheesemakers we have found that four times as much cheese is sold in December as in any other month. I don't know what the figure is for other cheesemakers, but I do know that we all end up with bare shelves in January.

Our customers have given us so much support, and as I reflect on our journey I can see clearly that it was our connection with the general public that nudged us firmly and unrelentingly in the direction of cow-with-calf dairy. When we launched those daily farm tours so many years ago we opened a window on our farming practice through which members of the public could look in. Perhaps more importantly, we could see their reaction to our farming methods, listen to their questions and, while doing so, question our own assumptions.

It was also our customers and the people who so wholeheartedly supported our crowdfunding campaign that were responsible for securing the final chunk of funding to build the new cheese dairy. Yes, after seeing the wave of public interest and support, the bank approved our loan. This was the final part of the complete farming system environmental transformation. It had taken us ten years to get here, but in many ways we were only just starting.

Chapter Thirty Three

David

As we embraced harmonisation with natural systems on our farm there was a noticeable change in the insect life. By changing our grazing management we gradually managed anthelmintic worm drenches out of our farming system, and I began to notice the prevalence of little holes in the cow pats, made by the unsung heroes of ecological farming – dung beetles.

It seems we now have at least three species of dung beetle on the farm. They are important for cycling the nutrients and carbon; breaking up the cow pats and burying a lot of organic matter in burrows in the soil. Their larvae are relatively large and attract attention from jackdaws, foxes and badgers who scatter the dung pats around the field in search of the juicy morsels. Unbeknown to us, when we had previously been dosing the livestock for parasitic worms, we had been killing these insects and disrupting these critical cycles.

Docks have always been a nuisance weed on the farm. I can remember getting paid sixpence a bag for picking up dock roots after the plough. By the time I returned home, technology had found the solution – weedkiller. A consequence of the weedkillers was that they killed just about every other weed – and herb – in the pastures as well. Farmers do not like to see docks in their fields, they're considered a sure sign of poor farming. I was really worried that once we had to stop spraying weedkillers, when we converted to organic 25 years ago, the place would be overrun with the things. I need not have worried. As so often seems to happen on this journey, nature came to our aid in the form of the green dock beetle and its larvae.

The green dock beetle and its little black grub of a larva, strip the greenery off the large dock-leaves and in dry weather the docks die from dehydration. So much so this past year that I actually felt sorry for the devastation brought onto the poor docks. They're not all bad. Their deep tap roots open up the ground and their leaves are high in mineral nutrients. They might be a wee bit unpalatable but when chopped into the silage they go down with the rest.

Bugs, beasties and butterflies. There's a load more of them about, and I know very little about them unless they impact on me as a farmer. Something else I've noticed and remarked upon to the farm team is head flies. Back in the day when we sprayed lots of pesticides about, after a spell of rain the head flies would eat you alive. They were a constant nuisance for the cows, making them irritable at the afternoon milking. We had to treat everything with insecticide to keep the animals from going demented. Well, no more. Sure, there are flies about, but nowhere near as bad as they used to be, and we never have to use insecticides to control them anymore.

Last, but not least, are the changes we are seeing in the animal kingdom. I mentioned earlier that we used to persecute predatory species on the farm – badgers, foxes, stoats, weasels, buzzards, kites, crows, etc. There is little doubt that this allowed some prey species to thrive – rabbits, hares, voles, and various ground-nesting birds like curlew, lapwing and oystercatcher. As social farm game shoots became less popular, along with certain species protection legislation, this persecution eased off and these predator animals began to re-colonise the area.

Natural biology is certainly very complex. You never quite know what is going to happen or why it happens, so my stories are our observations and our theories about what and why. The first thing we noticed was that as the badgers and buzzards increased, so the rabbits decreased. That makes sense. Initially I would see up to a dozen pairs of buzzards working across the gorse covered hillside. There were thousands of rabbits and there was plenty of food. Gradually the rabbit numbers declined until there were only a few dozen left on the entire farm. By this time there was only one resident pair of buzzards left and a visiting pair of red kites.

It wasn't just the buzzards that were having an impact on the rabbits. I'd often see a rabbit burrow dug out and fluff scattered around. The badgers were back, but something else was happening too. The numbers of adders had rocketed. There were adders in the visitor playground – they were climbing up the handrails, they were in the visitor centre, there was even the odd one that came into the farmhouse! We'd just scoop them into an empty four litre ice-cream tub and return them to the nature trails. I quite like adders. Wilma doesn't. But what was going on? As well as being the farm-problem-solving-builder, Jim is a part-time game keeper and is very nature aware. I asked him what he thought. 'It's the badgers.' he said. 'Ah', I thought, 'this would be the game keeper coming out – blame the badgers!'

'What do you mean, 'the badgers'?' I asked. 'Well, they've eaten all the hedgehogs,' he explained. That was certainly true. Hedgehogs used to be so common that we had to install little hedgehog ramps in the cattle grids to let them climb back out after they'd fallen in. As the badgers re-colonised the farm, you'd see hollowed-out hedgehog spiny skins all over. I've only seen one hedgehog on the farm in the past ten years, and it wasn't alive. 'Okay Jim, so what's that got to do with the snakes?' 'Well, hedgehogs can kill adders.' Really?! I had to check that one online. Sure enough, it was true, hedgehogs are resistant to adder venom.

In recent years there has been another change. Adder numbers have been declining. Was it their food source? There had been loads of voles in the newly planted woodlands, a food staple of adders. Had they eaten them all? Or maybe there was another explanation. Stoats were starting to colonise the farm. There was even a family established just outside the farmhouse garden where the hens live. If you hear a hen clucking after laying an egg, you have to be quick or one of the adult stoats will come zipping over the garden wall and quick as a flash, nip into the henhouse and disappear with the egg. Yes, stoats kill and eat adders.

Something else we've noticed is that the numbers of rooks have increased. They use the oak woods near the farm as their nesting rookery and would work the fields all winter looking for food. Back in the days when we persecuted them, we'd get grass fields that developed yellow patches in the spring. Sometimes this could be up to half the field. Close inspection would reveal the yellow grasses had been cut off at ground level. It was the larvae of the cranefly (daddy longlegs), also known as leatherjackets, and there were millions of them. The solution fifty-plus years ago was to spray the field with DDT. Later, when DDT was banned, we'd roll the field with a heavy roller to squash the little blighters, but they weren't called leatherjackets for nothing, and this had limited success.

What we've noticed in the past twenty, or more, years is that even when there have been warnings of a high risk of leatherjackets attack, we've never seen one. Sure, there are still craneflies about and, no doubt, their grubs, but not in the numbers that can cause serious damage to the grass, and I believe we have the rooks to thank for that.

Of course, there have been losers in this re-balancing of nature. The curlew, the lapwing and the oyster catcher have disappeared from the farm. They didn't have a hope once the badgers returned. A lot of the small birds have gone too. Perhaps that was caused by our attempts at scrub clearance plus the change of cropping from weedy turnips and oats to grassland. I don't know. When once it was common to see kestrels, sparrowhawks, the odd peregrine falcon and even an occasional hen harrier, it is only the occasional sparrowhawk now. Even the barn owls, who had little competition for the farm voles and hooted infuriatingly outside the bedroom window on crisp, clear winter nights, have declined, presumably due to the extra competition from badgers, stoats and adders for the poor old voles.

Before I leave this section on nature, I must mention the story of the swans. There are a pair of swans who hatch a clutch of cygnets down in Megan's Lochan at the bottom of the farm. Of the six or seven cygnets hatched they would be lucky to raise one or two to fledging. The nest was in a marshy end of the pond, in spring it was surrounded by water but as summer wore on the water dropped making it easy for predators to snatch the odd cygnet.

One spring we noticed one of the adult swans flying back and forth from this pond to another, nearly a mile away at the other end of the farm. Next thing we knew, the adults would be marching their week-old cygnets along the farm tracks towards the second pond which had a safer nesting spot. The problem was the three cattle grids along these tracks. Every year the cygnets would fall in, and it was a hairy exercise scooping them out as the adult swans hissed and flapped around above your head. That is until we put plywood sheets at the grids in preparation for that day. Naturally, that was the year they changed their route. They had figured out how to avoid the cattle grids.

Once again we saw one of the adult swans flying back and forth and next, they were marching through the visitor centre, down past the farmhouse, through the farm steading and up another track to avoid the grids. I was struck by the planning and sentiency of these beautiful and impressive birds.

Chapter Thirty Four

Wilma

With so much effort having gone in to designing a new system of dairy farming, we were determined that the cheese made from the precious cow-with-calf milk should be the most delicious, and the most nutritious, cheese possible. For us, that meant we were determined to make raw milk cheese. Unpasteurised cheese has a lot more character and taste, but more importantly, pasteurisation doesn't just kill off the bad bacteria, it kills off the good guys too, which means unpasteurised cheese is much better for your gut biome than cheese made from pasteurised milk. While unpasteurised, traditional farmhouse cheese used to be the norm, now there are only seven cheesemakers left in Scotland who make it. We are the new kid on the block.

Our first 18 months of cheesemaking using raw milk were relatively event free. We maintained good relationships with our Environmental Health Officers (EHOs) whose role had always appeared to us to be to assist as much as to enforce. That all changed after a tragic E. Coli O157:H7 case in Scotland in 2016. An unpasteurised cheesemaker, who had helped and advised us greatly, was blamed, even though E. Coli O157:H7 was never found in their cheese, and a court case a couple of years later found that all their controls were safe.

Our cheesemaking journey has been interesting and much more

complex than we could ever have imagined. Instead of the daunting enough task of simply learning how to be cheesemakers, all of a sudden we've had to become microbiologists and scientists. We've needed to learn to understand detailed scientific evidence into research on the physiochemical characteristics of cheese. Understanding 100+ pages of heavy scientific documents has become a pre-requisite for making unpasteurised cheese in Scotland as food safety guidelines have tightened.

It became the concern of many within the Scottish cheesemaking world that the authorities were trying to eradicate unpasteurised cheesemaking by stealth, by introducing new guidelines that were impossible to meet. Scotland already has the toughest laws in the UK on raw milk – the sale of it is banned completely – so the risk of unpasteurised cheesemaking being shut down was a very real prospect. All this was happening just as we were finalising plans for the new cheese dairy, the final step in our transformation that would mean we could make cheese in volume. The ramifications of a change in attitude to raw cheese production in Scotland would have an enormous impact on us.

In early 2019 the remaining producers of unpasteurised cheese in Scotland started a legal crowdfunding initiative to address this heavy hand of bureaucracy in court. We thought long and hard about whether to join this legal action, after all, the last thing any business wants to do is enter into legal proceedings, and especially not against a government agency with regulatory authority over your business, but we felt strongly that the legality of the new guidelines needed challenged.

Scotland's cheesemakers had been presented with a flow chart of criteria that all unpasteurised cheese made in Scotland should meet. That flow chart effectively meant that no matter which route you went down, the end result was the same – no unpasteurised cheese would ever pass 'go', and no new cheesemaker would ever be able to start making unpasteurised cheese. It could have meant that unpasteurised cheese from France or England could be sold and consumed in Scotland, but it couldn't be made here, effectively eradicating decades worth of knowledge and expertise from our nation's food heritage.

The authorities said we shouldn't be concerned about the 'guidance for enforcement' flow chart, it was simply guidance, it wasn't mandatory they said. We cheesemakers were still concerned. If there was a food poisoning incident where raw milk cheese was implicated and 'guidance' hadn't been followed, then the media and the authorities would have a field day.

Once again the general public showed their support for small scale businesses trying to do the right thing. The crowdfunding target to enable legal representation was achieved, and we felt a softening in attitude. It seemed to us that while the authorities might have the financial clout for a court case, there wasn't the political appetite for a public fight. Thankfully everything quietened down, an amicable understanding was reached and the dreaded flow chart has not been mentioned since.

We still have a very good relationship with our EHOs, but it saddens us, angers us even, that there's still no recognition by our government of the gut health benefits of nutrient-dense raw milk cheese.

Chapter Thirty Five

David

Sheep play an important part in our farming system, and we now farm them in quite an unusual way. The sheep enterprise produces lambs for meat, and the economic diversity they bring is no bad thing, but that's not the only reason we stock them.

The gut worm parasites of sheep are largely different to those of cattle and by alternating the grazing fields between cattle and sheep we have broken the life cycle of these parasites. This has cut to almost zero our need for veterinary treatments which, in hindsight, were having an adverse effect on insect wildlife. Now that we've eradicated the need for these treatments, we can see the natural biodiversity bouncing back. Another benefit of having sheep on the farm is that we can use them to graze young reseeded areas to control unwanted weeds. On the whole we quite like weeds on the farm, but it's all about balance and the sheep help with that. It works well.

Up until four years ago we farmed the flock in a fairly traditional way, lambing outdoors in March. We would buy Scottish Blackface ewe lambs from the same organic farmer each year and cross-breed these with a Bluefaced Leicester ram to produce what is known as the Scotch Mule, which comprised our main breeding flock. The Scotch Mule ewe made a very good and milky mother for the

Suffolk or Texel-cross lambs she produced. These lambs would grow rapidly and would be ready for market from four months of age onwards, and the lambs would be sold off the farm before the rams joined the ewes again in early October.

To get this level of performance took a fair amount of supplementary feed – around twenty tonnes of cereal and soya with minerals and vitamins each year. All organic of course, but in transitioning to a circular, sustainable food producing system, we also transitioned to 100% pasture fed farming. The supplementary feed had to go.

As the shepherd, I also had two important issues I needed to address. These Suffolk and Texel lambs undoubtedly grew fast, but the ewe often had to be assisted because the lambs were large at birth. This usually meant chasing and catching the ewe out in the field, which is no mean feat and it takes a good dog to help bring the ewe to a standstill. When I had to assist two or three lambings every day, that began to take a toll, and I certainly wasn't getting any younger.

The second issue was the toll this was taking on the ewes. I've heard that a ewe carrying and then rearing two good Texel lambs is the equivalent of the strain put on a high yielding Holstein dairy cow. Well, here we were with the low-stress, ethical dairy model for our cows but we were putting our ewes under huge stress to produce lambs. Agri-economists say that for efficiency reasons it is best to use a larger ram size to make the best of the ewe's potential. However, our experience with the cow-with-calf dairy experiment suggested that the economists weren't always right. Was there a better way?

Since returning home from Shetland I had been using Shetland rams to breed with our ewe lambs. The Shetland ram is half the size of a Texel, and the reason I used a smaller ram was so that the first pregnancy for the young ewe would be easy on her. I wondered what would happen if we used a Shetland ram across the whole flock? I'd noticed the young ewes would rarely need lambing assistance, so I figured using small Shetland rams should reduce the stress on the ewes, and reduce the interventions they'd need from me during lambing.

Which got me thinking, perhaps instead of lambing in March, could we lamb in May? As the ewes would be lambing onto fresh spring pastures we, perhaps, wouldn't need to feed them any supplements whatsoever. But lambing in May creates an issue because we wouldn't be able to shear the ewes until later in June. That leaves the ewes vulnerable to heat stress, getting itchy and becoming stuck on their backs as they try to scratch, which can be fatal. Their fleeces would also tend to get mucky from liquid poo from the young spring grass leading to blowfly strike, and fly strike is the stuff of nightmares.

So I wondered what would happen if we sheared them before lambing, in April? April can be a pretty chilly month in this part of Scotland, with nasty blustery showers. Would the shearing or the weather stress them and cause them to abort, or to get pneumonia? These were all pretty extreme changes. How would the sheep fare?

Previously we lambed all the ewes in one big field with lots of shelter. As they lambed, we would then move the ewes and lambs out into follow-on fields, then onto their spring pastures. It was a lot of work. In this new system we would be lambing straight onto the summer pastures and, if we knew which ewes had a single lamb and which had twins, we could match the quality of grass accordingly. This meant scanning in March, shearing in April and lambing in May. It was all a bit of a gamble, but I'd been thinking this through for several years and it was time to act.

The result? I wish I'd done it years earlier. Transitioning the sheep to this new system was much easier than transitioning the milking herd to cow-with-calf. We've gone from intense intervention during lambing to simple supervision. Rather than catching and assisting several ewes every day, I've only had to catch and lamb four ewes in four years, in fact we joke that the lambs hit the ground running. There is very low mortality, no mastitis and very few hoof problems, and there's also no mismothering now. This is mothering instinct gone wrong, when a sheep abandons one or both lambs after lambing, or when a sheep steals newly-lambed lambs from another ewe. It's all stress related and it just doesn't seem to happen anymore. I reckon the ewes are probably living a couple of years longer too.

There is one big change which some people might view as a problem, and that's the rate of growth, but I don't have an issue with it. Being from a small breed of ram, the lambs take quite a bit longer to reach market weight. Instead of four to six months they take seven to ten months. This means that I've had to reduce the numbers of breeding ewes so that there's enough grass over the winter months to sustain the lambs, but that's okay. The lambs are reaching the market at a time of year when lambs are scarce, so they're in demand.

Once again I've found that stepping back and facilitating a more natural, low-stress food system is paying dividends. We've almost completely stripped out cost, stress and welfare issues. The sheep are not getting any supplementary purchased feed at all, and very little in the way of veterinary interventions. But much more important than that, I can manage the whole enterprise myself on just a few hours a week, which gives me time for other things, and in the spring of 2019 I really needed that extra time.

The reason was that we were working on creating the cheese dairy that Andres from Andalucía had envisioned; transforming an ancient, semi-derelict threshing barn into the place where our cheese would be made, matured, cut and packed. The old barn had been built in 1787 and work began by removing and disposing of years of accumulated rubbish; putting up scaffolding and stripping off the old slate roof, and exposing the extensively rotten timbers. It turned out the whole lot could have collapsed at any time. Next, we took the old, two-foot thick, stone walls down to where the build was sound. This was fairly easy as the walls were built using a lime mortar which could be removed from the stone with a light chipping hammer allowing us to re-use the stone. In anticipation of the build, we had been preparing large containers of lime putty to use as a base for making lime mortar for the re-build.

We had built most of the modern buildings on the farm using cement and concrete blocks, so the lime mortar was a real departure, but we had done some reading about it. It dates back to Roman times and has a number of advantages over cement. Lime mortar never really goes rock hard like cement. This means even when there is settlement in the walls there are no cracks. Every wall we've ever built with blocks and cement has a crack in it. Another advantage is that lime mortar walls breathe. Condensation on them is rare as the moisture settling on them soaks in and travels through the wall to evaporate on the outside, which is a really good quality in a cheese store wall. Lastly, as we were seeing, the stone is re-usable as the old lime mortar is easily removed. We were starting to understand the benefits of harnessing natural processes in our approach to building, as well as our approach to farming.

Needless to say, Jim headed up the building team and by the end of the summer the derelict barn had been returned to its former glory with the internal walls and doors re-configured to meet its new function. Next it was a case of installing the new plant room, cutting and packing room, under-floor heating – which was to be powered by the anaerobic digester – concrete stairways, doorways, lift, wall linings, insulation, electrics, ventilation and extensive cheese shelving to take twenty tonnes of cheese. Then there was all the cheese-making equipment.

We completed the cheese dairy build in August 2019 and were in and making cheese even as the cheese store shelving was being built. Was it not ever thus? It was, and is, a beautiful building. On one of the walls is a list of names; a thank you to the contributors to our crowdfunding campaign whose generosity helped make it happen.

Chapter Thirty Six

Wilma

In late 2018 we began organisation for the crowdfunded Ethical Farming Conference in earnest. We've attended plenty of conferences over the years, but have never organised one. Where do you even start? Perhaps with a date. The third Thursday in May was chosen for all sorts of farming reasons. A marquee would be a good idea – how much does a marquee cost? Then there is the video and sound engineer and, of course, the food. Oh, and don't forget the speakers.

There were four farms involved in organising this event, all organic farms in the south of Scotland who are selling directly to the public. Chris and Denise from Peelham Farm, Pete Ritchie from Whitmuir Farm and Nourish Scotland and Bryce Cunningham from Mossgiel Farm in Ayrshire. Add David and me into that mix and you have a group of very busy, very strong minded individuals who are all used to being in charge. With the exception of Pete, I don't think any of us are practised at having to build a consensus. Agreeing on the speakers was the most difficult part by far.

We all had our favourites, but it seems that the regenerative farming sector is still a little too small. You could almost guarantee that one of us held a grudge against someone else's favourite. 'No way, he still owes me money' or 'he's too left field' or 'she's too mainstream'.

During the Christmas break David and I had been sitting looking at the list of proposed speakers and we didn't have agreement on a single person. We went to Pete for advice, and he suggested we listen to the proposals from others, take account of their objections, and go for a balance that David and I felt would work. This may be the first of many 'Ethical Farming Conferences' which could be held on farms around the UK, he counselled, but this one would be remembered as the conference at The Ethical Dairy.

By the end of February we more or less had our speakers. The two keynote speakers were Emry Birdwell and Deborah Clark from Texas who mob graze 5,000 cattle, which means moving them several times a day, and Dr Zoe Harcombe, a nutritionist who was vegetarian for many years but now advocates the health benefits of eating red meat. There were also eleven other inspirational speakers – from soil scientists to community veg growing activists – and there were optional sessions at lunchtime on topics including mental health and storytelling.

The star of the day was the weather, oh how lucky we were. There is something magical about the landscape of this part of Scotland in May, and on the day of the conference the farm and the surrounding countryside looked stunningly beautiful. If anything it was too hot, and it was definitely too bright.

Our first panic came the night before the event when the audio-visual company phoned me at the restaurant where we were entertaining the speakers. The marquee didn't have a dark lining, they said, and if we were going to be able to read any of the presentation slides on a sunny day, a dark lining was essential. No matter how much I insisted that a dark lining had been ordered, it was clear that it hadn't arrived. This problem filtered round our guests in the restaurant. One joked with David, 'Surely you've got enough silage wrap to cover the marquee?'

David tossed and turned all night. He was up at 5am planning how he would sort it, and at 7am he and a couple of the farm team were on ladders, up at the peak of the huge marquee, covering the entire 'speaker end' with the thick, black plastic sheeting normally used to line silage pits. The irony was this was supposed to be an entirely plastic-free conference! We had carefully designed plastic out of every single aspect of the day – from the catering to the name badges – and here we were wrapping half the marquee in the stuff! But it worked. The audience could now see the presentation screens and the plastic would definitely get reused.

The other star of the day was the food. There were so many people we are thankful to for contributing to such a spectacular spread of amazing food that catered to everyone's dietary choices; from keto-carnivores to vegans. Rude Health supplied the breakfast, Nick Green of The Green Butcher in Twickenham, who sells our beef and lamb, barbecued rose veal to perfection, our local bakery The Earth's Crust provided sourdough bread made with Scotland the Bread heritage flour. Most of all it was the Cream o' Galloway team who pulled out all the stops to produce and serve a fantastic array of salads and vegan dishes. I don't think I've ever been as proud of anything as I was of the food we served that day.

For David the day was just a blur. After sorting the shading of the marquee, he then led a total of six packed tractor-led farm tours throughout the day. We had livestreamed the whole event, and he ended up having to watch all the streamed conference sessions later on YouTube!

There was one big disappointment however, and that was the realisation about just how toxic we had become to the wider farming industry in Scotland. We genuinely believed that our keynote speakers were relevant to all livestock farmers in the south of Scotland. Emry and Deborah from Texas are down-to-earth large-scale farmers with 5,000 cattle that they farm in a way that has improved their soil and biodiversity immeasurably, while Zoe Harcombe is such an inspirational speaker and full of facts on the health benefits of eating red meat. Surely at a time when so many livestock farmers were feeling under attack, for health and climate reasons, a talk from Zoe would rally them? But no.

We approached NFUS (National Farmers Union of Scotland), QMS (Quality Meat Scotland), SRUC (Scotland's Rural Colleges) and AHDB Dairy (Agriculture and Horticulture Development Board) offering them our two speakers free of charge the night before the conference. We even struck a deal with a local venue who would only charge a token fee for refreshments.

Initially the staff at these organisations were positive, 'If we have people of that calibre in the area, then we should definitely use them' was one response. By pure coincidence the triennial industry gathering ScotGrass – a major grass based farming event – was being held just thirty miles away, the day before our conference. There would be thousands of farmers from all around the UK attending, so what could be better than some uplifting talks on the environmental and health benefits of turning grass into nourishment?

One by one these organisations all found reasons why they couldn't support it. The result was a unanimous 'no'. David was later told that the sticking point was the fact that both our speakers were promoting natural, 'grass-based' farming. No fertilisers, cereals or soya. For the movers and shakers in the Scottish farming industry, this was totally unacceptable. We now find all this quite comical, but at the time it was absolutely incomprehensible.

Almost without exception we find it's the new entrants – those just starting out in farming – who want to do things differently. Or it's the farmers who have had a career off the farm before returning home, many of whom are now retailing directly to the general public.

Putting the idea of a wider industry event behind us, we moved on to something more achievable – we'd end the conference with a ceilidh. We are so fortunate to have the best venue I know less than ten miles away from us – Laggan just outside Gatehouse of Fleet. It has an amazing view over the scenic Fleet Valley. 150 people enjoyed the delicious food and the band got us on our feet. The perfect ending to 24 hours that could not have gone better.

Chapter Thirty Seven

David

As we approached the end of the third year of our cow-with-calf experiment my confidence that this was the way ahead was surging. We had given ourselves a three year deadline to prove we could do it, or quit. After coming so far, I certainly wasn't about to quit.

That year, 2019, seems to have been something of a tipping point in how our system was perceived by others. No longer a joke, we were a threat to some, an inspiration to others and, perhaps, we were part of a solution for those who could see that the direction of travel for conventional dairy was heading nowhere good.

As well as the conference, we hosted several visits that year from international groups. This included a group of vets from Bologna in Italy, all very experienced with dairy cow management because they work with dairy farms supplying milk to Parmesan cheesemakers.

Then there were groups of researchers from the Agricultural Research University of Sweden in Uppsala, under the direction of Professor Sigrid Agenas. They were embarking on a research project to explore cow-with-calf dairy management, and the visit to see our system in action helped inform their research. This international interest was incredibly reassuring.

We decided long ago that our door would always be open to

anyone in farming who genuinely wants to know more about what we are doing. At times that open door has felt like an open invitation for a kick in the teeth, but there was a definite feeling at this time that things were changing. After twelve years of planning, testing, developing and refining this system we had learned a lot about our cows, calves and their environment. We'd also learned a lot about ourselves and how all these different factors play a crucial and closely interwoven part in creating a successful, low-stress, high-welfare, dairy system.

Somewhat out of the blue I was told I was a finalist in the CEVA Animal Welfare Awards. CEVA is a veterinary pharmaceutical company and their annual awards mark excellence in animal welfare. It turned out I'd been nominated by a veterinary lecturer and researcher. I went on to win their Farmer of the Year Award, and I was blown away to receive such a prestigious award.

In November autumn calving started once again. This marked the start of year four of cow-with-calf dairying, and we had an announcement to make. We had been so public with our cow-with-calf journey, and about our three year deadline, that the time had come to announce the outcome of our experiment. We issued a statement to a cynical industry to say that cow-with-calf dairy farming was viable at scale, and that we were committing to this system of farming permanently.

Earlier in the summer Ian, our designer, had filmed some video footage of a young cow – a first time mum – grooming her calf. The footage captured real tenderness and typical mother-calf grooming behaviour – the mum insistent on grooming, the calf resisting ever so slightly as she did so. By happy coincidence we realised that this cow was one of the very first calves to be born into our system, way back in the autumn of 2016. So, we used her experiences to tell the wider story of our farming system, and we introduced people to this beautiful young cow called One-Forty-Four.

People think that by identifying an animal by a number we are removing that animal's individuality. Well, for us at least, that's far from the truth. Sure, we could give them all names like Daisy or Esmerelda, but when you have 120 of them the names are quickly forgotten. That's not to say we don't have pet names for some of them, we do – but when you've been working with them from birth it's their character you relate to, not the number or name.

Cow 144's start in life was radically different from almost every other dairy cow in the world, in fact she's an important pioneer, yet she has no idea how special she is. For her, the way we do things at Rainton Farm is all she knows.

She was born on the 5th November 2016 and her mum was old 130, a steady, middle-of-the-road, Montbeliarde-cross cow. Her dad was a Swedish Red bull and she was conceived by artificial insemination. It's almost invariably the case that the calf of a cross-bred cow takes on the characteristics of the pure-bred father and, true to form, she looks like her dad.

She stayed with her mum, who we milked once-a-day, until early April 2017. For about six to eight weeks she was with her mum all the time and free to suckle on demand, but after that we began separating them overnight. They're not taken out of their mums' sight or hearing, just separated by a rail fence, it's almost like the calves have their own bedroom right in the middle of the dairy shed. The cows get a feed at night and then we tidy the cow beds and put the calves into their quarters while their mums are distracted.

The reason we separate the cows and calves is two-fold. Firstly, when the calf is young they can't drink all of their mum's milk, so we can be sure of getting some, but after a couple of months they can drink the lot. Separating overnight means we get at least some milk.

The other reason is that when the calf is with its mum 24/7, they don't eat solid food. Why would they, with milk on-tap whenever they want it? Overnight, when separated, the calf gets hungry and begins eating solids, and this is good for developing their digestive system. They also begin to build stronger peer-bonds during these overnight periods which helps greatly at weaning. This gradual transitioning process, at every stage of their lives, is a key part of the success of this system. We depend on the cow, the calf and us being happy and content. Making sure each and every change is introduced gradually reduces stress and improves the productivity and efficiency of this method of dairy farming.

When Cow 144 went out to grass in the spring she was fully weaned and she stayed with her female peer group. Back then, in year one, we didn't give the weaned calves any supplementary feed so they grew much more slowly than when they were suckling their mums. That was a mistake and we now give the young cattle a pellet made from lucerne.

When she was aged 18-20 months we introduced a young dairy bull to her group and nine months later, in March 2019, out popped a beautiful little calf. We spend quite a bit of time getting the young heifers used to the milking parlour before they calve. Giving birth is a big enough stress on its own, so if they know the parlour noises and the routines, that helps keep any stress to a minimum.

She entered the milking herd for the first time after her calf was born, and she is as good as gold. She reared her calf through the summer and they were fully separated in late August 2019. After weaning, we milk the cows twice a day for a couple of months because the calves had effectively been doing the second milking when they were suckling.

One-Forty-Four, one of the first cows to be born in our cow-with-calf system, is now one of our top cows. In fact, she and her peers are noticeably more productive than the previous year's intake. Was this because they were reared in the new system from birth and knew the ropes? I don't know. Maybe somebody will tell us someday. Her calf was a wee cracker too and has already joined her mum in the milking herd, which means we are now into our second generation of cows born into this cow-with-calf dairy system.

When we shared 144's story we announced confidently that this system works for the cows, the calves and for us. Getting to this point wasn't easy, but it was worth it. We believe this system can address pretty much all of the climate, welfare, environmental and ethical ammunition currently being launched at the dairy industry. Intensification is no solution to the food production challenges facing our planet; a kinder approach to dairy farming is the only way to go.

Chapter Thirty Eight

Wilma

The new cheese dairy meant we were able, for the very first time, to make cheese in large enough batch sizes to mean we'd have more cheese than we could sell directly ourselves. That's not to say there weren't teething problems. Cheesemaking is a craft, not a science, and moving our whole operation to a different building, with a different microclimate, working with bigger volumes and bigger machinery, was a steep learning curve. There were quite a few batches that had to be binned, and more than a few that ended up being labelled as 'cooking cheese' because the texture wasn't quite right. But we were on the right path, and we needed to find new outlets. One company was right at the top of our wish-list. We wanted Abel & Cole to sell our cheese.

We first met Abel & Cole at our conference; Chris and Denise of Peelham Farm were already suppliers and had invited them along. Our values and our approach were in synch, and within a few weeks we were in discussion with them about becoming suppliers. Chris and Denise had described them as one of their best customers in every sense, and so it has turned out to be for us. An exceptionally caring team who obviously want to do the best for their customers, but at the same time they look after their suppliers. They launched us with a fanfare to their customers. Their monthly 'cheese club' box of four cheeses normally includes one cheese each from four different cheesemakers. On the month we started working with them, they dedicated their whole cheese club box to us – it was quite the endorsement.

Since we launched Cream o' Galloway, way back in the early 1990s, we have become used to the cut and thrust of the commercial food industry. Margins are tight, suppliers are squeezed and lines can be dropped abruptly for reasons completely outside of your control. While I find it a rewarding industry to be in – producing products that people value and enjoy – it's not an easy industry, nor one that offers any get-rich-quick outcomes. So I find it important to work with people and companies with whom I have shared values.

When David and I first decided to start up in business together, he talked a lot about creating rural jobs, and keeping money in our rural community to help maintain public services like schools and doctors. Meanwhile, I just wanted out of the rat race. I knew that I wouldn't earn as much as I had done in Glasgow, but that wasn't why I was making such a major change.

To be honest David's aspiration of maintaining a rural community just didn't gel with me. I guess I was still in big corporation mode, and, to me, David's intentions were simply words that looked good in a business plan. I wince when I remember times during our start-up phase, before we launched the ice-cream, when I was still acting as if I was working for a multi-national corporation. In other words, I was a bit of a shit, making demands of local suppliers that they couldn't possibly hope to meet.

As soon as I employed our first few members of staff, that changed. I was now responsible for their security of income. There was no anonymous person higher up the ladder passing down a diktat that I could blame, it was all on me. The first time I had to lay off a member of staff, because there simply wasn't enough work through the winter, sat heavily with me and it made me explore how we could make the jobs here much less seasonal. In the short term that made the business less profitable, but the loyalty of experienced staff was more important than profit.

There is a vibrant community of small scale food producers here in Dumfries & Galloway. Many, like me, have moved into the region and will have been changed for the better by its people and its culture. Maintaining rural skills and artisan crafts are no longer just phrases used to tick a box in a business plan. This aspect of our business has become really important to me. I admire the skills of so many people working in the crafts sector, and boy do I marvel at Jim McKnight's ability to turn his hand to make and repair just about any piece of equipment we put in front of him.

The McKnight family have played a vital part in the developments at Rainton. Jim's father, Adam, worked for David's father, and his four other siblings have all worked here at one time or another. His sister Jane, now Jane Malcolm, joined Cream o' Galloway in our very first summer in 1992. She was born on the farm and she's worked with us for twenty years, with an eight year gap in the middle when she had her twins. Jane's skills are as valuable to Cream o' Galloway as Jim's are to the farm. I marvel at her tenacity and determination to find solutions to technical problems, but for me her greatest skills are the soft ones. Her ability to understand and empathise with the emotions of individuals in our team has always been important, but Covid added a level of stress to everyone that has made this quality even more essential. She has taught me a lot.

Over the last thirty years there are many people who have helped guide me from my previous corporate mindset to understanding how business can be a force for good. Barry Graham, a founder of Loch Arthur, a farm-based working community twenty-five miles away that includes men and women with learning difficulties, is one of them. Barry has influenced so many people with his integrity and vision, and he was instrumental in resurrecting traditional cheesemaking in Scotland. There are several deep links between Loch Arthur and Rainton, not least the old cheesemaking equipment. It had been gathering dust here in the 1980s when it found a new home, and a renewed purpose, with Barry at Loch Arthur. As I began my journey to bring cheesemaking back to our farm, Barry was always at the end of a phone with sage words of advice. More recently, when our wonderful visitor centre manager, Helen, decided it was time for a new challenge, she headed to Loch Arthur to manage their farm shop.

Another person who has inspired me over the years is Wendy Barrie, a chef who is an advocate for Slow Food and publishes the Scottish Food Guide. She will have no idea how often I quote her wise words. I first met Wendy about twenty years ago when she was advising businesses on how to improve their menus. We had been serving up the usual fare – soup, sandwiches and toasties with a wide variety of fillings – ham, cheese, egg, chicken and beef. All were organic, but our farm only produced the beef, we bought in everything else, and the poorest selling filling was our own beef. How could we make the beef on our menu more popular? Wendy's answer was not what I had expected. 'Drop them all,' she said, 'just offer your beef. If you don't have pigs and chickens on your farm, why would you offer it?'

While this was music to my ears it didn't feel like good business sense. The customers clearly preferred ham and chicken to beef, we were bound to get complaints if we dropped them. 'Change your menu, change your signage,' she said, 'make it much clearer that you're offering food from the farm.' She gave us the courage to try it, and it worked.

As a general rule, I find the bigger an organisation is, the more difficult it is to work with them. We have found working with supermarkets challenging in lots of ways. Price increases are refused and money-off promotions are demanded, and if the supermarkets refuse a price increase, then I feel we can't increase our price for the independent trade either. It would be very unfair if we increased our prices to the local deli, but not to the supermarket round the corner. So we've had several years where our costs have been rising, but not the price that we charge, and that causes real problems. We won't be the only ones.

Like so many people who have created purpose-driven businesses in the food sector, we try to respond directly to what customers ask of us. We know that some of our customers would like 'slaughter free' milk and there is a small charitable farm in England offering this, but we've heard the waiting list for produce is very long. This is not something we will ever be able to offer, and there are important reasons for that.

Fundamentally we want to demonstrate that cow-with-calf dairy farming can work for other dairy farms too, in fact we want it to become the norm. To do that means we need to operate our farm in a way that others can replicate, so that means no charity. There is nothing easy about these decisions, and when you work with these wonderful animals daily – caring for them, interacting with them and to all intents and purposes, becoming part of their herd – the hardest thing of all is sending them to slaughter to become food.

I am certain that if my life had taken a different path and I had never met David then I would now be vegan. In the early 1990s I would not have described myself as a foodie, but I was interested in animal welfare, and I 'did my bit', buying free range and organic food. I had two spells of being vegetarian, but long hours and a hectic city lifestyle meant I rarely had time to cook from scratch, so I wasn't particularly good at being a vegetarian. Looking back at that time, I would say I was a concerned citizen more than an informed one. I might have read an article in the Guardian, but I would not have looked for more detail.

However, I did meet David and consequently my knowledge of food production, livestock farming, environmental management, nutrition and climate change is markedly different to what it would otherwise have been. If I have one wish it would be that everyone ventured outside their own bubble and started to talk face-to-face about how we have all been duped by industrial corporations. Food production and farming are complex, highly politicised topics with huge vested interests. No-one wins in a broken food system except the multi-nationals. There is an urgent need to find other, better, ways of nourishing our population.

Chapter Thirty Nine

David

Now, in our fourth year of cow-with-calf dairying, we were, for the first time, no longer losing money. Everything was settling down and there was no turning back, but all the activity with the farm and cheese dairy had taken our focus, effort and our finances away from our original 'baby', Cream o' Galloway.

The visitor centre and adventure playground were both needing a lot of attention. The timber framed walkways and towers had been built by us twenty years previously and the repair costs were mounting. Our main insurer, who we'd been with since the start, decided to stop insuring playgrounds and we found it very difficult to find anyone to insure home-made play equipment, because the risks were so difficult for them to assess. Our premiums shot up by over £10,000.

In a final bid to breathe some life into the adventure playground, we built an electric go-kart track and bought a dozen e-karts from China. This proved popular and with the renewed interest we – well, I did, against Wilma's advice – decided to spend a serious amount of time and money over the winter of 2019/20 upgrading the playground. Of course, we all know what happened in March 2020, just as we were about to re-open our face-lifted visitor attraction. Covid-19. For the visitor centre and ice-cream business it was a case of damage limitation and, once again, survival.

As the world emerged slowly from the first lockdown in summer 2020 we got the go ahead to gradually reopen. Our adventure playground was mostly outside, and we did our very best to make it work, but it just didn't. In this new world, it couldn't. The playground had been built in such a way as to bring people together, but now we were forced – by law and by a responsibility to public health – to keep people apart.

Our wonderful wooden bouncy bridge and our most popular activity, aerial bouncy nets that we called Go Boing, were both designed to throw people about and to create unpredictable movement. There's no way you could maintain social distancing on those things. The flying fox, pedal karts and e-karts all required close-contact checking of harnesses to make sure our young visitors were strapped in safely. The high level of staff interaction needed to make everything safe and enjoyable meant queues, and any queue – even outside – was now a problem. Despite the government grants, for which we were immensely grateful, the visitor centre was haemorrhaging staff, time, energy and money.

On the other hand, for the cheese business, by marketing directly to our customers sales went stratospheric. Originally we had hoped we might get up to 20% of our cheeses being sold direct to the customer through our website. During 2020, it was 90%.

When Covid hit and the hospitality industry shut down virtually overnight, we realised we needed to increase our online sales dramatically. Luckily we had a website and a packing room that was already set up for direct selling, and we had marketing expertise to help us do it. We started running digital adverts on social media immediately. While the rest of the speciality cheese market in the UK stalled as wholesale markets collapsed, our sales surged, and our wonderful customers supported us every step of the way. At one point it seemed we were the only cheesemaker in the UK trying to manage problems with sell-outs rather than surplus stock. In fact, we were so close to selling out of our own cheese, that we started selling cheese from other Scottish producers who had lost their wholesale orders.

Looking back I do wonder whether it was our experience with Foot and Mouth Disease way back in 2001 that helped us respond to the crisis so quickly. It was certainly a steep learning curve for Wilma and her excellent cheese team. For me and the farm guys, little changed through Covid. We worked away in our own little farm bubble, the isolation suiting my character to a 'T'. I've confessed to Wilma that I'd quite like to be a hermit when I grow up.

Chapter Forty

Wilma

I feel almost guilty when I say out loud that we quite enjoyed the 2020 lockdown. It was certainly a time when most farmers realised just how lucky we are. We appreciated the joy of being able to work outdoors in what was one of the best springs we can remember. Maybe we just remember it because we could take the time to enjoy it.

The cheese team continued to work but I was the only person in the office. Our online cheese sales grew by a factor of ten during the lockdown period. My day still started before 7am as I met the team to give them the overnight online orders that needed cut and packed that day. After that, my farm job was to check two of the lambing fields. In the warmth of the spring sunshine it was just glorious.

In previous years I would spend the spring months rushing around checking every nook and cranny for ewes that were lambing, without taking time to even notice our beautiful setting. Normally I had to be back sitting in the office by 9am, when the phone would start ringing and I would rarely finish work before 7pm.

During lockdown everything was much more real for me. I could actually see spring springing into life. The fine weather made checking the fields a relaxing walk, with time to appreciate the wildlife, rather than a rushed chore. It was as if the whole world had slowed down giving me time to appreciate every moment, I started to notice every flower and hear every bird wherever I went.

Of course there were financial concerns for Cream o' Galloway, which relied on the tourism and hospitality industry, both for ice-cream sales and at our visitor centre. As time progressed it was becoming clear that there was major uncertainty around the business and while financial support from the government would see us through the short term, what should we do to make this business more resilient in the longer term? I distinctly remembered when David built the bouncy bridge that led to the adventure playground in the woods. 'This will last twenty years', he said. That was in 1996. It was now 2020.

Lockdown gave us time for a complete re-think of our strategy for Cream o' Galloway. This time it was me who was keener than David to fast-track changes. He was emotionally and financially invested in the adventure playground; the new e-karts alone had cost £25k. Were we really going to abandon something that was so popular?

At the crux of this decision was one simple question – what did we want to be doing in five years' time? We were in our mid-60s, we had to be realistic about our own energy and capacity, but The Ethical Dairy had re-kindled our enthusiasm. To be doing something that has so much support from so many people really gets you bouncing out of bed in the morning. On the other hand, managing an aging adventure playground that needs constant repair, especially when the general public had become increasingly litigious over those 24 years, had become a demoralising worry.

With that admission, it was easy to see what we should do, but for our existing regular customers it would be the most unpopular decision we have ever made. Making the announcement that the adventure playground was to close was one of the hardest things

I've ever had to do. It broke the hearts of many of our most loyal customers, and my heart broke a little too. But if I'm honest with myself, while Covid accelerated the decision, the closure of our beautiful, hand-built, woodland adventure playground had become inevitable as soon as its insurance costs had started to soar.

In the winter of 2020/21 we dismantled the indoor play areas and converted them into workshop rooms where we could hold hands-on events, like cheesemaking courses. Then, in the winter of 21/22, we started to dismantle the outdoor playground. We recognise that there will still be many people arriving at Cream o' Galloway with their children or grandchildren purely because of our adventure playground, but we just couldn't make it work anymore. Covid was the final straw that forced the end of David and his team's incredible playground.

Chapter Forty One

David

Cream o' Galloway is a very different place now. The massive ice-cream counter is still there, of course, but dismantling the playground marks a major change in what we provide to visitors. The outdoor bridges, walkways and the 50ft viewing tower have been kept and are being converted into woodland walks. Everything else – climbing, sliding, bouncing, crawling, running stuff – has been taken away. Crazy golf remains, and the e-kart tracks have been converted into a nine-hole crazy croquet course.

The farm and The Ethical Dairy brand is now much more visible, and consequently much more accessible to people. Farm tours that are suitable for families take place during school holidays, and I'm running regular in-depth farm tours covering soil health, pasture, biodiversity, carbon emissions and sequestration, anaerobic digestion, parasites and diseases, 100% pasture-fed, production without antibiotic, and, of course, cow-with-calf dairy.

I find leading these farm tours enormously rewarding. There is very little opportunity to talk to interested individuals within the farming industry because cow-with-calf dairying has, until very recently, been a no-go subject. Those who join my farm tours seem to be fascinated with our journey and with what we have discovered. Personally, I get a great deal from these events too, and the questions posed by our visitors are thoughtful and considered. What gives me hope is that we are now seeing increasing numbers of people from within the agricultural community booking onto these tours, curious about how we are making this work.

Meanwhile, back on the farm an unexpected problem reared its head. It was one I honestly never saw coming, but looking back, I guess this hard lesson was inevitable too. I'd been hanging on to a lot of older cows. We have a 'closed herd'. This is where the only cows that come into the herd are those that have been born on the farm, and the main reason for this approach is disease prevention. It has protected us from many of the infectious diseases that are endemic in the national herd, but means it takes much longer to expand the herd size.

Before introducing our cow-with-calf system, we had been carrying 100 cows and all the young stock to finished beef or breeding age. With the calves now growing at twice the speed, they were reaching market or breeding weights much earlier. That released a lot of forage and grass, which allowed us to feed extra cows. We figured we might be able to carry 30 cows more, which would help to offset the extra milk the calves were drinking. So our target milking herd size became 130 cows.

We aim to give the old faithful cows a final year of retirement, just suckling their calf. I realised that if I delayed retiring the older cows from the milking herd we could increase the herd numbers more quickly. That was a big mistake. It turned out that some of these old girls were carrying a nasty bug called staphylococcus aureus, which is highly infectious and causes mastitis. Not only that, but we were starting to pick this up in the milk and cheese tests.

Staph is complicated to manage, because testing an individual cow's milk doesn't prove anything. The disease is intermittent and it flares up from time-to-time, which means that a negative test is not confirmation that the cow is negative. The disease can be dormant, but then flare up at a later time. There are only two ways to get rid of the disease. Either through a heavy programme of antibiotic treatment, with no guarantee of staph eradication, or to cull the cows that are consistently showing high cell counts in their monthly milk tests.

We were now a herd that was certified free from antibiotic treatments, which meant that any cow receiving antibiotic had to leave the milking herd anyway. So, there was little choice, they had to go. Very sadly, the vet identified thirty high risk cows and we separated them from the rest of the herd. They had a wonderful final year, running outside with their calves until early December. We had no indoor space for them away from the main herd, and they were chronically diseased old cows, so when winter came I had to make an incredibly hard decision.

In an industry where the industrial cow barely reaches her third lactation, and the fact that our cows regularly go on to reach double figures in age, it does beg the question as to why longevity isn't deemed a measure of welfare by the dairy industry.

The industry's argument is that if lack of longevity is used as an indicator of poor management then farmers will be tempted to hang on to cows that should have been culled, say for chronically sore feet, or chronic mastitis. Supermarkets and milk processors offer dairy farmers direct supply contracts, usually at a price premium over the market average. In return for the price premium, they demand farmers meet or exceed certain standards on cow health and welfare. Health and welfare is assessed by the farmer recording incidents of disease, use of antibiotic, etc. and these are known as key performance indicators, KPIs. If a farmer exceeds any of these KPIs then the contract will be under threat and that could cost the farm a six figure loss of income, assuming they can even find an alternative buyer for their milk.

I remember a vet joking that a customer of theirs with 2,000 dairy cows got less twisted stomachs in his herd than we got. I was astounded. Out of 120 cows, at that time, we maybe got one every five years when the industry average was one or two twisted stomachs for every hundred cows per year. I expressed my disbelief and the vet replied, 'we have a zero tolerance of twisted stomachs.' That was like saying you have a zero tolerance of catching a cold, it made no sense at all!

I let it pass and later asked Charles – because he had experience of these kinds of things – what the vet had meant. He replied, 'Twisted stomachs are a KPI.' It took a moment for the penny to drop. 'So they kill them?!' 'Sure. Longevity is not a KPI.'

Longevity has always been important to me. I want our herd to live long, healthy, natural lives. The decision I had to make that winter was, in the end, the decision that minimised the health risk to the wider herd, but boy it wasn't easy. Several of these cows had been faithful servants for many years and you can't help but become attached.

Another problem we ran into was that we were carrying far too much stock. The bull cattle had been going as eight to ten month rose veal into food service, but food service had all but closed down over 2020. Nobody wanted bulls over twelve months of age, so, we had to keep them until they were big enough to go as finished beef cattle at 20-24 months. In March 2021 I took the decision that we would have to sell all the young stock that were not suckling their mothers to another organic farm. By sheer chance, we entered the summer of 2021 with a much-depleted herd of cattle. Surprisingly this worked out for the best because it turned out to be one of the driest summers we have ever experienced in this area. From late February to mid-September instead of about twenty-five inches of rain, we got five and a half. We only just had enough grass to nourish the herd.

In contrast, the second half of September brought wet, but good grass growing conditions that allowed the milking cows to continue to graze outside until the end of November. We've never done that before. Mostly the milkers are inside by the beginning of November. It was very weird weather and I guess these changing weather patterns are a sign of things to come.

The good news is that the milk is now completely clear of staph and the herd is at a very low risk of mastitis. By the end of this year we will be back up to 120 cows, 130 cows by 2023, and the cows and calves are bouncing with good health and wellbeing.

Chapter Forty Two

Wilma

Just before Covid I had a cold that never quite went away. Nothing serious, I just found myself clearing my throat several times a day. I tried steam inhalation and herbal remedies, but nothing worked. I assumed that when summer came it would go away. It didn't.

Like everyone else, I wasn't going near a doctors' surgery during Covid, so it was April 2021 before I made an appointment with my GP. Her thinking was that since I wasn't a smoker and since this had been going on for 18 months, it was unlikely to be lung cancer, but she'd arrange an x-ray just to check it out. Up until this point, cancer hadn't even crossed my mind.

David

By the spring of 2021 Wilma's months of persistent throat clearing had developed into a cough. Her GP referred her for a scan with the reassuring statement that if it was a lung cancer that had been untreated for 18 months, then she'd be dead already.

Over the course of the summer Wilma went through a series of scans, biopsies and tests but they found nothing. In early July she suffered breathlessness, began to feel unwell and by August was feeling some chest pain when lying on her right side.

A scan in late August revealed that her right lung was three quarters full of fluid. She was taken into hospital immediately, put on a heavy course of antibiotic and a catheter was inserted into the lung to drain the fluid. She had been suffering from something referred to as 'walking pneumonia', a form of pneumonia where the sufferer still feels healthy enough to keep going.

After a week of treatment Wilma bounced back in great form. So, it had been this low-grade pneumonia all along, we thought, but the tests continued. In August she had a PET scan and biopsy. By now they had picked up indicators of cancerous cells. These may have been remnants of breast cancers from decades earlier, or it could be lung cancer. The former was treatable, the latter wasn't.

Wilma

We had a meeting with the oncologist who explained I had lung cancer and proposed a course of chemotherapy and immunotherapy drugs. This was just incomprehensible, I'd already had a double mastectomy and as far as I was concerned I was going to be like my mother and live cancer-free until I was 90. I try to suppress the 'why me?' question and, like most people these days, I used Dr Google to find out more.

Something I initially thought might explain it is that women who've had breast cancer and received radiotherapy have an increased risk of primary lung cancer many years later. So the 'why me?' question may well be answered by the fact that almost thirty years earlier I had radiotherapy that scarred my lung, and the only way to come to terms with that is to appreciate the thirty fulfilling years since the first cancer diagnosis. Later discussions with the oncologist suggested that the cause of my tumour was unlikely to be the earlier radiotherapy, so we'll never know. In the end, the cause doesn't matter.

To be honest I think I'm still in denial. They told me that the tumour was 10cm. I was astonished, and really couldn't understand how it had taken so many procedures and medical investigations to prove malignancy in a tumour that large. I don't think I'll ever understand it, but denial is a pretty good place to be at the moment.

David

I guess I was in some kind of denial because, in my head, the difficulties of detecting any cancer whatsoever, the repeated message that if it was lung cancer Wilma would be at death's door already, the diagnosis of a chronic pneumonia and her bounce back to great form convinced me it was all done and dusted. Wilma was tough. She was unsinkable. I therefore sat in a state of shock as, in early-October, the oncologist delivered the bombshell.

I can't remember much of the detail other than twelve-month life expectancy, up to twelve-month extension with treatment and something about palliative care.

We sat in the car in the hospital car park and wept.

Wilma had stage-four lung cancer. The main tumour was four inches in size but there was evidence that the cancer had spread outside the lung. They put her on a pretty aggressive, twelve-week treatment programme of chemotherapy with scans at six-weekly intervals to assess the response.

The oncologist explained that surgery and radiotherapy were ruled out, and the best we could expect from chemotherapy would be that the spread of the cancer might be halted.

At six weeks, the scan showed a 50% reduction in tumour size. At twelve weeks it had shrunk a further 25%. The oncologist was delighted and put Wilma onto a much less aggressive treatment regime. She then extended Wilma's predicted life expectancy to two to five years, which should have been shocking, but from where we had come, it felt like great news.

Wilma

Fortunately I still feel very well. I've been having treatment every three weeks for almost a year. The tumour has shrunk considerably and is now stable. Thankfully the treatment isn't affecting me too badly. I'm a little more tired than normal, but that's about it.

I still start the day by checking three fields of sheep which takes about 90 minutes. I certainly don't wake up every morning thinking about cancer. There have, in fact, been some benefits. David and I have stopped moaning at each other, and our social life has greatly improved as we tour the country visiting friends and reassuring them that I'm still going strong. As most people with a serious illness will say, can you please just treat me as normal? I still want to be me. I still want to work hard and I want to have fun. However, I do need to plan.

Normally people given a diagnosis like this are concerned about their children and grandchildren, they want to see them secure in life. I don't have children. David's three children are all in their forties now and content with their lives, but I do regard The Ethical Dairy as my toddler and Cream o' Galloway as the child that won't leave home.

I like to think I've got another five years, while David, ever the optimist, says he's sure I've got another ten. I can't expose our businesses to an uncertain future, so my focus now is making sure that those businesses – and the jobs they sustain – are secure.

The one person I do worry about is David. I've always said to him that I will go first and that, actually, I want to die first, but I warn him that he can't become a farm accident statistic. Now more than ever, I need him.

None of us know what's ahead and while we all say we want to die suddenly in our sleep, for loved ones, that's the worst possible way to go. At least this way we have lots of time to love and laugh, and appreciate the attraction we first felt thirty years ago. And if David is right and I get another ten years, then I can enjoy my seventies by visiting our customers and leading tours of the farm and cheese dairy. I look forward to that.

Chapter Forty Three

Wilma

While our personal world was falling apart, opportunities to speak about our approach to farming were opening up all over the place. COP26 was due to take place in Glasgow in November 2021 and we had been involved in discussions with three different organisations about having some form of presence at COP.

Pete Ritchie of Nourish Scotland had helped organise our conference and he was involved in a number of different COP projects. His goal was to create a farm in a park in Glasgow, near to the SEC where COP was being held, and he wanted us to bring cows and calves to the park. Another organisation wanted us to have a cow and calf in the SEC itself for the whole two weeks with thousands of people going past them every day. Ahem, nope. That was never going to happen. There was also talk of us appearing in special COP editions of BBC radio programmes. If our small experience was anything to go by, there must have been hundreds of projects being proposed and still being worked on in the final couple of months before the event.

Clem Sandison, a freelance artist and regenerative farming expert, was co-ordinating the farm in the city park project. We'd known Clem for years and four weeks before the start of COP we phoned to ask her opinion on how likely the farm project was to go ahead. It transpired that this was just a small part of a much bigger project being run by a consortium of environmental charities, and the event now had the working title of 'Cows in a Net Zero World'. Clem's original idea had been to bring some cows and calves to a former municipal golf course in the east end of Glasgow to show how urban land could be used differently. She wanted to demonstrate agroforestry – cows grazing amongst trees in a city park – in action, with the cows' grazing integrated naturally into an urban setting. That original site was deemed too far from the action so another park, closer to main COP26 venue, was identified.

We travelled up to Glasgow and met Clem at the park to do a risk assessment. Loading and unloading the two cows and calves was the major concern. Cattle careering along Argyle Street in the centre of Glasgow might get a lot of publicity, but not for the right reasons. We checked on parking restrictions for the COP period and had long debates about the practicalities of it all, but in the end it all came to nothing. The park was under the control of the UN for the duration of COP, and the UN said no to anything happening in that park due to their security policies. The agroforestry demo didn't happen, but 'Cows in a Net Zero World' went ahead as a discussion event at Nourish Scotland's COP26 Hub, with David participating in a debate that included farmers from the global south. It was a fascinating discussion, and that approach, while less high profile, was infinitely less stressful than taking cows and calves into the centre of Glasgow.

All that time spent in Glasgow also coincided with another occasion. We have been fortunate over the years to have won a number of business awards, but the VIBES awards have always been important to us. They're Scotland's environmental business awards and Cream o' Galloway won it in 2007, back in the days when the environment didn't really feature strongly in most companies' corporate objectives. In 2021, COP26 made these awards more important than ever before.

Initially our written entry won us a 'Good Practice' award and got us through to the shortlist for an Outstanding Achievement award. The day of our interview with the judges, I was at the hospital getting another procedure to determine whether I had cancer or not, so David needed do it on his own. I was actually relieved at this, because David and I can work against each other in this type of situation. I focus on giving all the facts succinctly while David enthuses at length, I nip his knee under the table to tell him he's talking too much, one of us rolls our eyes and we both go off our stride. David had it all his own way that day and I know he would have been outstanding when explaining the environmental achievements on the farm and in our cheese dairy.

He picked me up from hospital later that afternoon and told me that the judges had well-tuned antennae to spot any greenwashing. He thought it had gone well, but we both felt that a livestock farm would not fit neatly into the lead up to COP26, and expected that alone to exclude us from contention. So we were astonished to be told a few weeks later that we had won an Outstanding Achievement award. The organisers asked us to produce a short video for use at the online ceremony, and everything needed to be kept under wraps until then. In the meantime, shit happened and the lung cancer diagnosis was confirmed. Chemotherapy was to start in just over a week. In fact it started the day following the VIBES award ceremony. Bitter-sweet indeed.

The actual ceremony was humbling. All the shortlisted businesses were outstanding. Business has changed. So many business leaders genuinely do want to make the environment the focus of their activity, and it's incredibly rewarding for us, personally, to be part of that movement.

Once again, thirty years after the first time, I was having chemo treatments every third Wednesday, just as we were heading towards

Christmas, our busiest time of year. The team around me were exceptional. Of course I still wanted to be in control, because that's just me, but I was banned. 'Let me handle the paperwork, I promise not to interfere in all the cutting and packing,' I said, but no, they'd worked that out too. There was a slow realisation beginning to dawn that I, like everyone, am not indispensable.

Chapter Forty Four

David

Our ethical dairy farming system isn't just about letting calves stay with their mum, it's about so much more than that. Our goal is the creation of a circular, regenerative food production system, fed by sunlight and rain, that has a positive environmental impact, delivers nutritious food, provides good quality jobs and works to the highest possible standards of animal welfare. It might not have been easy, but our farm is living proof that it can be done.

Taking part in COP26 events was interesting, but frustrating too because change is simply not happening fast enough. I know that we're a pain in the ass to those in the industry pushing an intensification agenda, and I get it, I really do, but I didn't expect to feel like I was pushing these issues too hard at COP. Change is difficult, hugely challenging and it's very risky, but the world is changing too and there is no turning back for any of us. Farming needs to change and it needs to change urgently, faster than any of us could ever have imagined. We started changing 25 years ago and we're still not done. Our cow-with-calf dairy farming system isn't perfect, not by a long shot, there's still lots to do.

One of the biggest challenges that remains to be resolved is the final stage for the animals we raise. For a cow that has lived her whole life, maybe 18 years in one place – her home – transporting her many miles by road to slaughter is unbearably cruel. At the present time there is simply no alternative, but there should be an alternative.

On-farm slaughter and mobile abattoir is the final stage of this journey towards a dairy farming transformation that I feel would be complete. Could it happen? I hope so. The Scottish Government commissioned research into it a few years ago, and that research concluded it was viable. It would take commitment from government and the industry to give it a try, but I know of many farmers who are deeply unhappy with the current 'options', or lack thereof. The livestock we rear deserve a better end than they currently get. I will continue to speak up about the need for this kind of infrastructure.

As I look ahead I see so many warning signs of the fragility and unsustainability of our food system in the UK. The series of financial shocks starting with the recession of 2008, followed by a decade of austerity, which achieved very little in terms of shoring up the country's finances. Hot on the disruption of Brexit came the pandemic which has left us in an even more exposed position in terms of guaranteeing our ability to acquire adequate supplies of commodities. The war in the Ukraine and the sky-high price of energy are accelerating inflation to levels we haven't seen for decades.

I remember back in the seventies it was deemed prudent to build an interest rate of twelve percent into farm budgets. In those days, for those of my father's generation, debt was viewed as something to avoid. Now debt is the norm. We are facing a perfect storm of economic catastrophe. With over a third of our food coming from imports, it's time we re-assessed our food policy from the perspective of global instability and our weakened financial strength.

I suspect that in the near future, rather than genuinely sustainable changes to food production, there will be calls to increase UK food production through intensification, alongside rewilding of much of Scotland's upland landscapes. It will be called 'sustainable intensification', not because it's particularly sustainable but because chucking in the word sustainable will make it sound more palatable. It will still require fertilisers and pesticides but, we will be assured, these will be applied in greatly reduced quantities based on soil and crop needs, with the aid of satellite navigation and artificial intelligence. Precision agriculture is the bedfellow of sustainable intensification. Policy people, the supply industry and most farmers get excited by this prospect, but it's basically business-as-usual, with a techno tweak.

Technology is great when it works, but it's expensive to buy and run, which means this approach favours the bigger operators who can spread the costs over more product. Larger scale requires bigger machinery, and bigger machinery needs more room – bigger buildings, bigger fields, less hedging, fewer ditches... This approach still needs fertilisers and pesticides, and we know that no matter how many assurances we are given, much of these end up causing long-term environmental damage.

While precision agriculture requires less chemical input, it still requires input to be available. The intensive use of technology also requires support for that technology to be available in the event of a breakdown, otherwise, everything grinds to a halt. That technical expertise is unlikely to be on our doorstep. Furthermore, with the speed of technological advances, within a matter of a few years, that technology often becomes obsolete. This long and very specialist supply chain leads inevitably to fragility and a high risk of failure.

Finally, this technological dependent, large-scale production system leads to a reduction in, and de-skilling of, the sharp-end workforce. The high-tech skilled jobs are based in the urban centres, while the mundane jobs are filled by the on-site, rural workforce. Problem identification is done through computer-based monitoring systems and specialists are pulled in as needed. The crafts of soil, crop and animal husbandry are gradually lost, and with it the loss of meaningfulness and job satisfaction. This is all happening now and it's really not good, but there is another way.

What I think should happen is a rapid acceleration towards a way of farming where natural systems – based on the interactions between animals, plants and microbes that have been fine-tuned over millions of years – harness the power of the sun to produce an abundance of food for all. All we need to do is better understand these processes so that we can facilitate them, minimise their disruption and harvest their products.

Sound like a fairy story? Well maybe I'm living in fantasy land, because this is where I am. At each step of my journey the experts have said, 'It can't be done!'

Well, you know something? We're doing it. And you know something else? Once these natural systems got going, they're as productive as when we were using all that toxic and environmentally destructive stuff to which our industry has become so dangerously addicted.

As far as our farm is concerned it doesn't matter that fertiliser, feedstuffs, pesticide and even energy prices are going through the roof, because we use very little of them. Our nature-based food system will continue to produce food, so long as the sun continues to shine and the rain continues, in moderation, to fall.

For the sake of life on our planet and the nourishment of our people, it's time politicians and policy makers stopped believing the fantasy of 'this time will be different' techno-fixes and started to harness the staggering potential of nature.

Chapter Forty Five

Wilma

Just before Christmas, I got an email from Marjorie in the office. 'Have had a call from a gentleman called Carlo from the BBC Panorama programme. They are looking to do a programme on the dairy industry. He wondered if this was something he could talk to David/you about?'

To say we were reluctant to have anything to do with it is an understatement. The reason? The fall-out from BBC Scotland's *The Dark Side of Dairy* programme several years ago still affected us.

Back then we were, perhaps, naïve in understanding what taking part in a programme like that involves. Speaking to the interviewer is only the start, because what follows is a feeling of powerlessness over how the story will be told, and which parts of the interview will make the final edit. So when Panorama phoned, David and I said to each other 'this has a Dark Side of Dairy ring about it' and we ruled out taking part there and then. We ignored the call. Carlo phoned again. I picked up the call this time. I told him I'd speak with David, but that we really didn't want anything to do with it. He was persistent and, eventually, he changed our minds.

We spoke to the production team at length, shared all of our concerns, explained our previous experience and we asked an awful lot of questions. We hoped there would be an opportunity to talk about soil carbon and ecology as well as animal welfare, but just because we get excited about soil carbon sequestration, doesn't mean that it makes particularly good TV.

The BBC Panorama team visited our farm in early January and as soon as they showed us footage filmed at a dairy farm in Wales showing acts of violence against cows – and recorded our reaction to it – we understood where this programme was going. I asked them to stop because I simply couldn't watch any more. The programme was broadcast in mid-February and I was happy that my contributions were left on the cutting room floor.

The previous month we'd been mentioned in passing on BBC One's Dragon's Den, and our website almost melted with the volume of digital traffic as prime time viewers searched for us online. We knew we needed to be prepared for the immediate impact of an appearance on Panorama. Rachel, our web developer, spoke with the server company to discuss the technical changes that would be needed to keep everything stable should we experience another surge in website visitors.

On the night of the broadcast David and I were watching the TV, with me monitoring our website, meanwhile Lorna was looking at social media comments and Rachel was keeping an eye on our web server. At the first appearance of David, as the presenter mentioned The Ethical Dairy, our website crashed. Despite everything we had put in place the server was struggling to cope with thousands of people landing on the website simultaneously. Fortunately the server company was on standby and they managed to get everything back up and running within a few minutes. Meanwhile comments on our social media channels exploded. There were so many questions, most of them people asking where they could buy our milk, and the simple answer was that they couldn't.

The first thing we did was publish posts encouraging people to visit the Cow Calf Dairies website, an online list of all the cow-with-calf dairies in the UK. There aren't many of us – most of the dairies farming this way are small – but for some of the people looking for cow-with-calf milk, our friends in other parts of the UK were an option.

In the week following the broadcast, what was just the same as before were the attacks on our social media accounts by vegan fundamentalists, and my use of the term fundamentalist here is really important. I have openly said that, had life taken a different path for me, then I would likely be vegan myself, and I know that for many people adopting a vegan lifestyle is a sign of concerned citizens wanting to make a difference. However, we had to ban a few people from our Facebook page that week, and the reason was that a few activists had stopped attacking us, and had started attacking members of the public who were interacting with our posts.

We know that dairy farming can be a polarising issue. We understand people have very strong views, and we know a programme like Panorama can lead to people wanting to express their views and their emotional response to what has been screened, but the reaction to Panorama was unlike anything else we have experienced.

It's easy to pass judgement on the actions of another person, and how could anyone ever excuse someone who hits a cow in the face with a shovel, as was shown on TV that night? But the people behaving in that way are as much a victim of an inhumane system of food production as the cows. They work unacceptably long hours in a highly pressured, deadline-driven system that has, perhaps, made them no longer capable of understanding how cruel their behaviour is. We have always said it is the system that is the main problem with dairy, not necessarily the individual people within that system. That is why we want to see a complete system change in the dairy industry, a transformation that improves the working conditions of people, as well as the welfare of animals. The only way that change will happen is for people to demand it. It seems that's what people are starting to do.

Over the past few months we have been inundated with questions. The public response to that Panorama programme is one of the main reasons we decided it was time to tell the full story behind our cow-with-calf farming system. Those social media questions 'if The Ethical Dairy can do it, why can't everyone?'

The answer is very simple – because it's really, really difficult. But it can be done, it needs to be done, and government policy could make the UK a global leader if they were to incentivise this approach.

We often say that those within the agricultural sector who visit us to find out more about our system are nearly always new entrants. There are also a few like David and me, where a partner arrives on the scene from outside farming and is shocked by the existing system.

What has been heartening post-Panorama is that 'real farmers' have started to visit. Most often it's those farmers who are selling directly to the public. Their customers are beginning to ask whether they follow a cow-with-calf system, so those farmers want to find out about what's involved in doing so. Recently we've had visits from farms bigger than our own. They ask whether our system is scalable to their farm. We tell them it won't be easy, but yes, we think our cow-with-calf dairy system is scalable to most family-sized farms.

I'm really proud of what David has achieved here, what we've both achieved I guess. And while there have been very difficult and distressing times during the past ten years, we've been able to support each other emotionally to keep at it and to find solutions. There are always solutions. We both know better than to assume this is the end of the journey, but the system now feels right, and it's working well for us.

To the best of our knowledge we are the biggest dairy farm in Europe leaving calves with their mothers to suckle. It's a stand out claim.

It would be even better if more farms convert to cow-with-calf dairy and someone knocks us off our throne. We don't want to be exceptional or special. We want a dairy farming revolution that leads to cow-with-calf dairy farming becoming normal. That is now the end goal.

Chapter Forty Six

David

Following a path of intensification has marched our whole industry into the crosshairs of an oncoming climate war. It's a path I refuse to walk.

Farming – beef and dairy farming in particular – is now widely perceived to be one of the worst perpetrators of environmental degradation and climate catastrophe. During the course of our lifetimes, agriculture has changed from being the valued means of nourishing a population to one of the primary causes of runaway global warming. It doesn't have to be this way, but sticking our heads in the sand in the hope it all blows over is just not going to work.

Contrary to powerful, vested interests who are demonising all beef and dairy, grass-based beef and dairy can deliver good health and positive environmental outcomes. That is particularly true of landscapes like ours in south west Scotland. This country, with its natural advantages for grass-based livestock systems, has a real opportunity to rethink our whole farming culture.

Government and farmers like us should be natural allies in furthering a future approach to farming that works with nature; pioneering a pasture-based, regenerative approach that is as sustainable as it is productive. If our industry's representatives stick with their current agenda of technology-based, greenwashed 'solutions' and intensification in a country that excels at pasture based systems, it will rightly place our whole industry in the frame as a health and climate liability.

As I reflect on human nature and our reluctance to change, I think of Percy Bysshe Shelley's sonnet Ozymandias. In the poem a traveller happens upon the remains of a statue in the middle of a vast desert, engraved on which are the words 'Look on my Works, ye Mighty, and despair!'

I imagine that once this statue was part of a magnificent palace overlooking fields of grain, lush pastures, woodlands and great herds of grazing animals, tended by well-fed citizens of that kingdom. Now it was all just desert.

Of course this is just imaginary, yet soil scientists, archaeologists and climatologists have found that this has in fact happened and is still happening. They warn that land management practices of the past have created much of the desert and arid land we see today.

Our soils are only one strand in the intricate web of life on our precious blue planet. The degradation of those soils impacts every other strand of life – biodiversity, flooding, drought, food security, climate change, ocean acidification, pollution, you name it. We need to change our systems, our methods and our lifestyles, and we need to change fast.

We don't have all the answers, and we don't have everything right, but we think we've figured out some of the pieces of some of the answers, and here's what we know for sure.

1. Net zero, agroecological food production can be done viably, profitably, and at scale.
2. Resorting to the quick-fix, sticking-plaster approach so favoured by the industry must be resisted. It didn't work before and it won't work now.
3. Governments and policy makers need to incentivise farming

system transformation to regenerative, agroecological practices immediately.
4. They need to listen to and draw on the experience of those who have already done it.
5. This needs to happen today.

To those annoyingly repeated accusations about the incompatibility of profitably feeding the world and environmentally friendly, carbon neutrality, what I would say is this. As far as we're concerned, feeding the population of the world in a way that doesn't wreck the planet is far from incompatible with viability. If we are to address our climate catastrophe, doing so should be our number one focus. But I'd go further.

As the costs of conventional dairy farming rocket, I find that our costs of production are coming into alignment. The large consultancy firms are predicting that the average cost of milk production will exceed 45 pence per litre in late 2022. I estimate that by the end of 2023 that will equal our cost of production. Not only that, but the conventional costs of production tend to rise over time, whereas ours will continue to fall as we improve our productivity, which we will. This is a productive, affordable system.

This has huge implications for the profitability of conventional farming, where prices for farm products tend to lag behind jumps in the costs of production. Sudden falling profits can lead to falls in production as producers cut back on the stuff they've become dependent on. Falling production can quickly lead to food shortages and price spikes for consumers, resulting in a cost-of-living crisis.

Can a cow-with-calf, agroecological dairying system 'feed the world'? There are several aspects to this question. First off, does the existing intensive dairy model feed the world? If we look at the amount of dried food, in the form of cereals and soya currently used to produce one tonne of dried milk food value, it works out that we need about 2.5 tonnes of dried food value to produce each tonne of dried milk. Even once we have adjusted the volume of food in, to an equivalent measure of food out, we find that for each tonne of milk food value produced from intensive dairy, there is a net loss to the world of around one tonne of food for human consumption. Then there is the societal cost of deforestation to produce the soya, and the biodiversity loss from the removal of hedges and wetlands to replace them with heavily fertilised monocrop ryegrass.

To collect, process and deliver to the retailer the cow-with-calf, regeneratively produced milk from up to twenty small farms, versus a single 2,000-cow industrial farm, could add about ten pence per litre to the shelf price, assuming we achieve a similar cost of production per litre, but think of the benefits! Benefits to the environment from more biodiverse soils, pastures and all the insects and animals feeding off that; less pollution from soluble fertilisers, pesticides and antibiotic entering our water systems, rivers and oceans. Benefits to society from a more resilient, less volatile food supply and price; fewer costs for removing fertiliser and pesticides from our drinking water; more well-paid, skilled jobs for local families in a socially valued industry producing safe, quality food for our nation. And last but not least, benefits to our farm animals from a longer, happier and healthier life, all for just ten pence extra per litre.

We will speak with any policy maker, academic, journalist, politician, scientist, environmentalist or influencer about our experience. We will gladly share what we have done and what's made a difference to us. It took us ten years to change our farming system, because the transformation of complex systems takes time. We are now almost out of time. This is a crisis of our own making. We need to un-make it. Every day we lose through inaction takes us a step closer to the tipping point of irreversible climate catastrophe. It's not too late if we act now.

For both of us this has been a fascinating, sometimes exhilarating, often very testing, but extremely rewarding life journey. Our practical experiences have demonstrated clearly to us that we humans are a critical part of an incredibly complex system of life that we meddle with at our peril. It has convinced us that environmental, social, welfare and economic sustainability is not just interconnected – it is one and the same thing.

We will continue to push at the perceived limits of dairy farming practices, and we will continue to share what we learn for as long as we are able to do so. We would wholeheartedly like to thank everyone who has joined us on this journey at every level, giving us the moral and financial support to go on. It has allowed us to glimpse a roadmap to a better place for all life on our planet. It has given us hope.

The Ethical Dairy System

We get a lot of questions from people who want to know more about the detail of our farming system. For many members of the public the desire for a deeper understanding is driven by a need to know whether our farming practice fits with their own ethical principles and values. We understand that everyone has their own boundaries of what they consider ethical or not, and we respect that.

We also get a lot of questions from people within the farming industry. Often these are questions from former dairy farmers – people who converted their dairy farms to beef or who walked away from farming completely. Many of these people became disillusioned with the conventional dairy industry, just as we did, and are curious about how we have made our system work.

Sometimes, increasingly now, we are being asked about the detail of our farming system by dairy farmers who are interesting in trying cow-with-calf dairy for themselves. For anyone serious about piloting cow-with-calf dairy we strongly recommend a visit to Rainton Farm, to see first-hand how it works.

For everyone who is curious about how we do it, here is a summary of our farming system.

A Guide To Cow-with-Calf Dairying

By David Finlay

Cow-with-calf dairying is still very much in its infancy and this guide is just a description of how the system works for us in the context of our farm and its various resources.

I personally see cow-with-calf dairying as part of a larger, whole-farm regenerative approach that tries to address the environmental, social, ethical and economic challenges facing the dairy industry at this time.

I also see cow-with-calf dairying as an opportunity for the small scale, high welfare, family dairy farm to clearly differentiate their dairy products and access a premium market with little further marketing. The image of a dairy cow and her calf is a powerful one, and this is a market the big players will find almost impossible to access.

1. Before You Start

Are you really sure you want to do this? This is not an easy option, at least not the way we do it. Cow and calf welfare is the priority. If you are weaning at three months or earlier, in our experience, the welfare of the calf will be heavily compromised and the cows will become distressed, with implications for their own health and wellbeing.

Do the people who are working with you on this system change actually want to make it work? If not, it won't work. This is a system that relies on a stress-free environment for the cows to let down their milk and for the calves to thrive.

As dairy farmers, we are indoctrinated to believe that the next litre of milk is the most profitable. A dairy calf in a conventional rearing system will get around 250 litres of a formula milk over an eight-week period. Organic calves get around 420 litres of your organic milk over twelve weeks. In our system the calves are drinking 2-2,500 litres of our organic milk over an average of five months. Are you prepared to accept that?

Before making up your mind, go visit some of the guys doing this method of dairying. Better still, offer to spend a few days at a busy time, calving perhaps, working alongside their team. Seeing and doing is believing, and it can be inspirational.

2. The Herd

We had a traditional herd of Ayrshire cows until we began planning this project fifteen years ago. We wanted to breed a robust cow that could look after herself, even when we might get things slightly wrong. She also had to produce a calf that would be suitable for the beef market as well as provide us with milk. She would need to have good feet and legs for walking to and from the pastures, and have a good natural resistance to disease, particularly mastitis.

I always swore we would never have Holsteins on this farm because they were, generally, designed for the industrial model of farming, but this is not entirely true. The Holstein stable has examples of bulls for all types of production. At the time I had become disillusioned with the Ayrshire breeders who seemed to be obsessed with 'show type' – they looked good but didn't perform.

We ended up with a three-way cross. Robust, low yielding but high solids Holstein; reasonably milky but good beef and temperament (important!) Montbeliard; and strains of Ayrshires that went to the Nordic countries a hundred years ago – the Viking Red.

These breeds are 'threeways crossed' so that each bull breed used is two generations away from the same breed. This ensures a good level of hybrid vigour, as we know the mongrel is naturally more fertile and healthier than the pure-bred. It also gives the best compromise between milk and beef production. We use artificial insemination and sexed semen because the dairy bull calves don't meet our need to have a finished animal for the rose veal market at eight to ten months.

Effectively we have two block-calving herds. One group calving from mid-October to mid-December, and one in April to May. We make ice-cream and cheese from our milk, so milk solids are important. One of the issues that arises from milking cows that are suckling their calves is the let-down of butterfat, or rather, the lack of it.

Our cows without calves produce milk with an average butterfat of about 4.8%. Those that are suckling give us milk with an average 1.8% butterfat. This can vary quite a bit. The cows who are more prepared to share their milk with us tend to give us higher butterfat milk. These also tend to be the more relaxed and confident cows. I guess if they are relaxed in the parlour, they are more likely to release the hormone oxytocin which causes the tiny sachets in the udder (alveoli) to release the high butterfat milk. Apparently, it is the same for humans.

By having the two herds calving six months apart the low butterfat milk from the sucklers is balanced by the high butterfat of the weaned cows, to average around 4% butterfat, which is fine for cheese-making.

We had thought that the cows from the Holstein bull would give us most milk, because that is what they are primarily bred for, but that has not been the case. What we've found is that temperament is probably a bigger factor than potential yield. This is because higher yielding genetics seems to be associated with more nervous cows. If a cow is nervous and easily stressed, in those first five months when she is feeding her calf, she won't release milk in the parlour. If we don't get that milk in the first five months, we will never get it. But it is important not to jump to judgement as a poor early sharer can rise to become a big hitter once into her third and fourth lactation.

Next, you will need a bull. If you are using only artificial insemination the effect of suckling on the delay in cows returning to oestrus after calving is to extend that period by about 35 days. This is known as lactational anoestrous. As we are looking to have a tight calving block, this is disastrous, but if you introduce a bull, his presence, actions and pheromones reverse this delay. Of course, if you don't want progeny from that bull, he will have to be vasectomised. This works well for us, and we have absolutely no problems with fertility in our suckling cows.

3. Managing The Herd

For cow-with-calf dairying to work, we believe it must be operated in an environment that is stress-free. For cows that are suckling to let down their milk for us, they need to be relaxed. There is a minimum of banging and crashing of gates and machinery. No shouting and absolutely no physical abuse of the animals. Sticks are banned.

Each stage of the cycle of life is gradual and stress free. Starting with heifers approaching calving for their first time, they are about to endure the hugely stressful experience of producing a calf. Then we will expect them to come into the milking parlour, a completely alien environment, and accept a milking machine being attached to their teats, and not object?

We bring the heifers into the parlour (after milking the cows) several times a week for a few weeks before they calve. This gets them familiar with that environment and the sort of handling they will experience after calving. It works very well, and we rarely get any issues. It's also safer for us in the parlour. It is noticeable that this calm approach has made it safer to walk among the cows and calves and there is little chance of being kicked. It has also significantly reduced the number of scuffles between cows as they sort out the pecking order when re-entering the herd after calving. Visitors comment on the overall calmness of the herd.

Our calving boxes are twelve feet by twelve feet. They each have a self-fill water trough and a restraining gate. We bring the cows into the calving boxes as they begin calving. Bring them in too early and they can hang around for days in the relative comfort, effectively bed blocking. We have CCTV linked to our phones so we can monitor both the calving boxes and the area for the near-calving cows.

The dry cows are kept on a long fibre, low quality forage right up to calving. This reduces the risk of LDAs – twisted stomachs – post calving. It also stops the calves from getting too big and makes for easier calving, which means less stress for the cow, and it stops the cows from getting too fat as fatty liver syndrome can be a cow killer. In the last fortnight before calving they also get a kilo of straight lucerne pellets each day as they will be getting a couple of kilos of these a day post calving. We don't feed to yield. Everybody gets the same, though sometimes we give the co-operating cows a wee bonus. We don't use any mineral or vitamin supplements, nutritionally the only extra are the salt licks.

The cow calves. If necessary, we will put her behind the restraining gate in the calving pen and assist, but this is unusual.

We check everything is ok and treat the calf's navel with iodine. We then select a three-litre sachet of frozen colostrum from the freezer and place it in the defrosting bath. This bath has a rotating sachet holder in it and the water temperature is held at 40°C. No hotter or the antibodies will be damaged, and the calf's protection will be compromised.

Supplementing the cow's colostrum with this good quality frozen colostrum, collected from the first milking of older cows, has been an absolute life saver for all calves born indoors. Since introducing this colostrum management into our calving protocols, we have not seen clinical signs of the killer disease, cryptosporidium. Crypto was something of a nightmare in our first winter of cow-with-calf dairying and colostrum supplementation is an essential within the first couple of hours.

We give the cow a 20-litre bucket of warm water (the post-natal 'cup of tea') and once she has had a chance to lick the calf, she is lured behind the restraining gate with a bucket containing some lucerne pellets. They do like them. At this point the calf is attempting to get on their feet and the instinct to suckle is most powerful. A few hours later the calf is much more aware of its surroundings and what is 'right'. Us intervening and helping it find their mother's teat, if this has not already happened, is often resisted and this can lead to a downward spiral as the calf associates the teat with a negative experience.

Often a dairy cow will have a low udder relative to the calf and the calf might struggle to find the teat. In the old days this didn't matter too much as we could give the calf colostrum by stomach tube, but now it's essential that the calf knows where the milk is kept and can find it easily. Restraining the cow and distracting her with some lucerne pellets allows us to remove sealant in the cow's teats (which we use on all cows at drying off) as this can stop a calf from getting any milk.

Even before the calf is on its feet, we get the calf suckling. Three hundred sucks is the goal. This ensures that the calf gets a good feed from its own mum but also by this point it knows where the grub is kept. Next, we collect the three litres of colostrum from the water bath and pour it into a bottle connected to the stomach tube. We slip the stomach tube down the calf's throat and allow the colostrum to gurgle into the calf's stomach. They don't like this much, but if I skip this step I know that the calf is at risk, and it could put others at risk too.

Within a couple of hours of calving the cow and her calf are off to a great start. We leave them together in the calving box for around 24 hours to recover and bond, then we take the cow into the parlour for her first milking. Sometimes the cow will object to being separated from her calf at this point, but that's not a problem. We have an auto-tandem parlour where each cow has her own compartment, and there is enough room for the calf to stand alongside the cow while she is being milked. This calms the cow, and she will let down her milk.

After this, we leave the gate into the calving box open, allowing the cow and calf to mix with other newly calved cows and their calves for three or four days before entering the main herd. This gives the calf time to learn about mixing with other animals and to be able to find its own mum in a social group. We then move this small group of newly calved cows and their calves into the main herd.

At this point and for the next two or three months, the calf will have 24 hour access to its mum. We will milk the cow once-a-day. When indoors, the calf will have access to an area that the cows can't enter called the calf creep. This area is very popular as the calves get a bit older and begin to look for a quiet area away from the cows. They tend to hang back here when we bring the cows in for milking in the mornings, and during the summer months when we bring the cows in for milking, the calves peel off into their creep area as the cows wait to come into the collecting yard. They understand that it's their 'room' in the dairy shed.

Once the calves reach about three months of age they can drink twenty litres of milk, and more. At this time their mothers' milk production is also beginning to decline which means our share is rapidly reducing. Around this point, the calves need to be separated from their mothers overnight or we get no milk at all.

If the calves are indoors, they and their mums are moved to the 'big-calves' side of the barn where their creep area has cubicle divisions similar to those of their mothers. They also have their own grooming brush – electrically operated vertical and horizontal rotating brushes that are activated by the calf lifting the brushes with its head. The calves have seen these in the cow areas and quickly learn how to operate them. They love their brushes.

Separating the calves from the cows in the evening, indoors, is pretty simple. The cows are given a feed of about 1kg of lucerne pellets down their feed fence and we brush-up their cubicle areas. As we brush, we move the calves along and shut gates as we go, and as we finish brushing all the calves are in their creep. The gate is shut for the night and it's all very relaxed and straightforward.

Outside is a bit more awkward. We have fenced tracks running between day and night fields. The trick is for one person (or a good dog) to bring the cows and calves along the track. Halfway along is an opening into the calves' creep paddock with a low bar to prevent the cows going in.

A second person stands on the track on the opposite side from the creep entrance. As the cows and calves come past the creep entrance, that person edges forward as a calf approaches and the calf dives through the creep gate. The cows walk on by and into their night field. I know it sounds unlikely but it's actually pretty quick, simple and effective. Interestingly if a calf manages to dodge past and into the cows' field, they will almost inevitably return to the creep gate to join their pals. It takes about thirty minutes to bring sixty cows and their calves out of their day field, along the track, separate them and give them their night snack of lucerne pellets as a reward.

We avoid using the same calf paddock for a second year and graze sheep on it to avoid any stomach worm build up. We will use it again in the third year. The calf creep paddocks also have a shelter for the calves in bad weather.

Overnight separation has other advantages in addition to the extra milk we get. It gets the calf used to the physical separation, and it starts to break the cow-calf bond and strengthen the peer bond. Also, as the calf gets hungry overnight, it eats a significant amount of solid food, preparing their rumen for full weaning. We noticed that with *ad lib* milk from their mums, calves would eat very little solid food. Why would they, after all? But this would mean the growth check at weaning, and thus the stress, might be enough to trigger disease.

The final stage is full weaning. About ten days before we leave the creep gate shut in the morning, and apply Quiet Wean nose tabs to the calves. These plastic tabs just take a few seconds to apply and dangle from the bit of skin dividing their nostrils. There is no skin punctured. They are pretty ingenious.

The tab is about two inches deep and four inches wide and hangs in front of the calf's upper lip. They allow the calf to eat solid food and drink water but as the calf reaches for their mother's teat the tab pushes the teat away. It's very gentle and very clever, albeit very frustrating for the calf, and for their mothers, who groom the calves a lot to encourage drinking. We keep the routines the same over this period. Then on the ninth or tenth morning we don't open the creep gate. There is a bit of grumbling, but nothing compared to the uproar we used to get before we discovered these tabs. Next morning, we remove the tabs.

At this point the calves are drinking around fifteen litres a day, so we move the cows on to twice-a-day milking for two or three months otherwise we will lose a fair proportion of those fifteen litres, and it eases the pressure on the cows' udders. As the cows approach the last couple of months of lactation, we move them back to once-a-day milking. Effectively we will be milking cows once-a-day for half the year. It's an easier life for everyone and in winter this conveniently coincides with the festive period.

April weaned calves stay in the barn for a couple of weeks after the cows start going out to grass to get them used to not seeing their mums. We then turn them out to a nearby paddock because they like to see the barn and their mothers going about for another week or so. Then they are ready to move away to their summer pastures. We've found that if we rush this move away from their 'home' territory they get into a bit of a panic, and we can find some back at the farm the next morning. So, gently does it. Also, it's good if there are a couple of old, 'retired' cows with their calves who can join the weaned calves to show them the ropes.

Similarly, the September weaned calves stay in their creep paddock for a couple of weeks after weaning and are then moved to silage aftermaths. The calves continue to get a kilo of lucerne nuts which allows us to better monitor their diet and thus health. No in-feed minerals or vitamins are fed, but we have found that we need to give the calves a copper bolus at weaning, which gets them through this growth stage.

As the cows come to the end of their lactation and approach the next calving, they would normally be given a minimum of 60 days 'holiday' – the dry period. Having discussed this with a research vet who was studying this system of dairy, we decided that, due to the low demands of this system on the cow, it would be ok to cut their holiday period to a minimum of 40 days, and the cows seem happy with this.

4. The Infrastructure

The parlour. We opted for a five-a-side auto-tandem parlour because any other type would have required us to separate cows and calves prior to milking. Auto-tandem is well suited to this type of dairying. The cows enter a side-passage one at a time and then enter one of five possible individual milking cubicles with pneumatically operated gates controlled from the parlour pit.

The cow stands alongside the milking parlour pit and can see everything that's going on around her. The milking machine cluster is attached to the cow's teats from in front of the cow's legs, not from behind as in every other milking parlour style, and the cow has a lot of room.

Calving boxes. The cows will need to be restrained to allow any calving assistance, clearing teat canals of teat sealant and assisting the calf to find the teats and have a good suckle. It also keeps the mother out of the way when giving the calf its colostrum top-up by stomach tube. We find that a small, hinged gate the shape of a cantilever cubicle, to allow access to the cow's udder with a fixing chain onto the calving box dividing gate does the job.

As our waste system is 100% liquid, we have a rubber mat covering the entire floor of the box with a four inch sponge mat under the rubber cover on the upper half of the box for lying comfort. We use a dust free sawdust to keep the beds dry. There is a 5% slope on the floor of each box to allow liquid drainage. There is a self-fill water trough in the corner.

There are twelve calving boxes for around 60 cows block-calving over a two month period. The boxes are cleaned and refreshed twice a day. They lead onto a tractor-scraped passage which is scraped once a day. This passage leads to a feed fence where the group of newly calved cows can self-feed silage.

5. The Economics

This was one of the hardest things to get my head around. How could a dairy system where the calf was drinking over 2,000 litres of our organic milk ever be a financially viable proposition? Before we went organic a calf got 4.5 litres a day of formula milk for eight weeks, about 250 litres. When we went organic, we had to feed five litres a day for twelve weeks. That's 420 litres of organic, whole milk which, believe me, I grudged. Now the calves would be drinking more milk than they'd ever be worth, it was crazy! But it works, and here's how.

Once we got our new disease and nutritional challenges sorted out, we found that the suckling was stimulating extra milk production. In an average 100% forage dairy system the expected milk yield is about 4-4,500 litres per cow per year. If you are organic you will have to feed at least 420 litres of that milk to the calf, leaving about 3,600-4,100 litres for sale. In 2021, our cows gave us an average of 3,000 litres and we know, with a bit of selective breeding, we can get that up to 3,500 litres within a couple of years.

In addition, the calves are growing more than twice as fast in that first six months than they did before. We are selling most of the bull calves into the rose veal market at eight to ten months at an average liveweight of 350-400kg. This is around 16 months earlier than before, but at 150kg lighter. The dairy heifer replacements are coming into the herd about eight months earlier.

The forage spared by getting a quicker turnaround of the young stock has allowed us to increase cow numbers from 100 to 125-130 without increasing our stocking rate. So the loss in value of the beef cattle sales is made up for by having 25 more to sell, and the milk sales which would have been around 410,000 litres a year will now reach around 450,000 litres.

Acknowledgments

Most authors acknowledge the support of their long suffering partners in this section of a book. Since we are each other's long suffering partners, that makes writing this part challenging. However, coming to the end of the book writing process will bring some changes for us, mainly because we have very different body clocks. Unlike most farmers, David isn't at his best in the mornings, so almost all his book writing was done late in the evening. Meanwhile Wilma likes the peace and quiet of the morning and gets up early to do anything that needs a clear head. We look forward to synchronising our time zones again.

Many people have helped us write this book. First and foremost is Lorna Young who badgered us for years before we committed, and once we had, she helped us make it happen. Then there is Ian Findlay who created the artwork for the book cover. Ian has done our design, photography and videos for years, and we were cock-a-hoop to discover he had experience of helping people self-publish books.

There are many early readers to thank. Friends and family members saw the first draft, along with a few people who didn't really know us, but whose interest in reading and knowledge of books made their feedback invaluable. Gillian Khosla, Carol Carr, Ann Campbell, Ann Barclay, Wendy Barrie, Orla Shortall, Clem

Sandison, Helen Fenby, Christel van Raaij, Julia Hamilton, Lorna Brown, Mark, Margaret and Christine Finlay – your encouragement and your sharp eyes for our typos was greatly appreciated. Your kindness in sharing constructive criticism, comments and corrections helped improve the book immeasurably. From your feedback there was sufficient consensus to understand the changes needed, and it also gave us encouragement that our writing had the makings of a book.

We cannot thank Karen Campbell, Helen Browning, Sue Lawrence and Pete Ritchie enough for your generosity of time and the wonderful reviews you gave after reading a near-final draft of the book. You undoubtedly gave us reassurance that we had written a book worth reading.

Many of our early readers and those mentioned in the book volunteered to be proof readers. It was wonderful – almost like crowd sourcing the proof reading. Special mention goes to Hilary Hawker and Cathy Agnew who were last minute and very thorough proof readers.

We are deeply thankful to everyone who has supported the crowdfunding campaign to self-publish this book. Returning to Crowdfunder four years after our first, transformative experience with this platform felt like a closing of the circle, and it seemed like a pragmatic way to sell the book too. We are so grateful that interest in this book saw us hit the funding goal to print it in less than a week.

Love and thanks in particular go to our book Sponsor, Neil Kerr, whom we suspect was an early doubter but is now a vigorous supporter. Thank you Neil.

We are so fortunate to have Wigtown, Scotland's Book Town, on our doorstep. A big thank you to Wigtown Book Festival for adding us to the 2022 festival programme – by doing so you gave us a deadline to get this book finished, which we needed!

Over the years we have had important advice and moral support from researchers, vets and suppliers. Professor Sigrid Agenas from the Swedish University of Agricultural Sciences was the first academic to visit our cow-with-calf system. The fact that she then sent her whole team to visit us afterwards was a huge confidence booster – thank you Sigrid. And thanks also to Professor David Logue and to his student Tom Harris, whose report on our cow-with-calf trial challenged us to have another go. Remember how you said, 'This is a bomb waiting to go off' when you first visited us, David? How right you were.

Our vet practice also deserves special mention. We've always had a tongue-in-cheek relationship with them – after all, how many more mad-hatter ideas could we throw at them? But when the chips are down they always take us seriously and advise us of the best treatment and management approach. They have been instrumental in helping us get the bugs out of the system.

South of Scotland Enterprise have recently encouraged and supported us to explain the totality of the innovation we have developed over the years; offering advice and understanding, as well as funding. This book is one element of a broader project that seeks to record and make accessible to others everything we have learned. Our sincere thanks to them.

This whole farm transformation could not have happened without financial support from the EU and government organisations over many years. The Scottish Government, LEADER, Dumfries & Galloway Council, Scottish Enterprise and South of Scotland Enterprise have all helped us to transform our farming system and our business. As have loans from friends and family members which saw us through financial troughs, acting as both carrots and sticks to keep going to make the transition work.

We are immensely grateful to customers, visitors, those who contributed to our 2018 Crowdfunder and to everyone who follows and interacts with us on social media. It is difficult to express how valuable your emails, letters and spoken words of support were in keeping our spirits up during some of the darker periods of this journey. Thank you so, so much.

And last, but by no means least, a huge thank you to all the members of staff who have worked here over the years. You have been fundamental to everything we have achieved, whether you were immersed in all the farm changes, or focused on keeping the existing show on the road, we simply could not have done any of this without you. Thank you from the bottom of our hearts.

David & Wilma

Book Sponsor

The production of this book is generously sponsored by

Neil David Kerr

Supporters

The production of this book has been generously supported by the following people.

Jannette Bevan
Sara Dixon
Sean Earnshaw
Giuseppe Falco
Kim Fligelstone
Liz Guenther
Louise Henry
Deborah Hill
Gillian Khosla
Jane Mackay
Tammy Mak
Kieren Mildwaters
Siobhan O'Donnell
David Peter Norris
Jennifer Reid
Jo Scorah
GJ Simpson
Jessica Elizabeth Stokes
Claire Wotherspoon

And Karen Wallwork in memory of her husband Ron, who died 10 August 2022

Making a Difference

The need to transform our food systems is urgent. The following organisations have all helped and supported us over the years, and continue to make a difference in driving change.

Compassion in World Farming
www.ciwf.org.uk

Nourish Scotland
www.nourishscotland.org

Pasture For Life
www.pastureforlife.org

Soil Association
www.soilassociation.org

Sustainable Food Trust
www.sustainablefoodtrust.org